Guide to JCT Minor Works Building Contract 2016

JCT Minor Works Building Contract (MW)

JCT Minor Works Building Contract with contractor's design (MWD)

RIBA Publishing

Sarah Lupton

Guide to JCT Minor Works Building Contract 2016

Reprinted 2018

Published by RIBA Publishing
66 Portland Place, London, W1N 1NT

ISBN 978 1 85946 638 4, 978 1 85946 782 4 (PDF)

British Library Cataloguing-in-Publication Data
A catalogue record for this book is available from the British Library.

JCT Intermediate Building Contract
At the time of writing this Guide, the 2016 edition of the JCT Intermediate Building Contract was in development. Any references to that form are therefore to the 2011 edition.

Commissioning Editor: Fay Gibbons
Project Editor: Alasdair Deas
Designed and Typeset by Academic + Technical, Bristol, UK
Printed and bound by Page Bros, Norwich, UK
Cover design: Kneath Associates
Cover image: Shutterstock: www.shutterstock.com

www.ribapublishing.com

Foreword

The Minor Works Building Contract is by far the most widely used standard form of building contract and plays a vitally important part in the procurement of small-scale building projects. The 2016 edition of the Minor Works Building Contract comes in two versions: one that includes provision for the contractor to carry out some of the design, and one that does not. While this new edition employs the familiar logical layout, clear format and simplicity, its attractive brevity means it is inevitably more dependent upon implied terms than some other JCT contracts. For example, the procedural rules are minimal and, for those unfamiliar with contract administration and law, this may raise questions.

Sarah Lupton's new *Guide to MW16*, which follows on from her excellent *Guide to MW11*, offers comprehensive guidance and does so in language that is easily understood. Organised by themes, the book is a straightforward analysis of the contract in the light of today's legal and practice landscape, referring to recent case law and clearly distinguishing that guidance which applies only to the With Contractor's Design version. The hard-pressed practitioner also will find the Guide helpful as it outlines the changes from the 2011 edition. Practitioners will be particularly pleased to see the useful indexes and will doubtless come to depend on being able to dip quickly into the book for specific help during the course of a project.

I would thoroughly recommend the book to both architecture and other construction students on the threshold of undertaking their professional examinations. The comprehensive up-to-date coverage clearly and succinctly exposes the legal ramifications of the contract. Sarah Lupton's rare combination of being a legally trained architect who also runs the MA in Professional Studies at Cardiff University makes this book the ideal student companion.

Neil Gower, Solicitor
Chief Executive, The Joint Contracts Tribunal
September 2016

About the author

Professor Sarah Lupton MA, DipArch, LLM, RIBA, CArb is a partner in Lupton Stellakis and directs the MA in Professional Studies at the Welsh School of Architecture. She is dual qualified as an architect and as a lawyer. She lectures widely on subjects relating to construction law, and is the author of many books including this series on JCT contracts, the *Guide to the RIBA Domestic and Concise Building Contracts*, *Which Contract?* and the 5th edition of *Cornes and Lupton's Design Liability in the Construction Industry*. She contributes regularly to the International Construction Law Review and acts as an arbitrator, adjudicator and expert witness in construction disputes. Sarah is also chair of the CIC's Liability Panel and the CIC Liability Champion.

Contents

About this Guide

The JCT Minor Works Building Contract was first published in 1968, and since that date it has been used extensively on small to medium-sized building projects. Its continuing popularity derives from its relative shortness and simplicity, together with its long and secure track record. It is undoubtedly an appropriate choice for smaller projects, where the additional features of longer forms are not needed. Despite its reduced length, it still includes a reasonable amount of flexibility, with the MWD version providing for contractor design input, and the optional supplemental provisions allowing the user to add features such as collaborative working, performance indicators and negotiation.

The 2016 edition will no doubt contribute to and maintain its popularity. The drafting has been simplified and rationalised in several areas, including the certification provisions and the clauses covering action following damage to the works. Insurance Option C has been modified to allow for more flexible solutions to insurance where work is done to existing buildings. In addition, provisions have been incorporated relating to fair payment, the Public Contracts Regulations 2015 and the Freedom of Information Act 2000. This makes the contract an ideal choice for smaller public sector projects, such as enabling works contracts, particularly in situations where it is intended that a JCT form may also be used on a related, larger scheme.

The Guide assumes no prior knowledge of the form. After a general introduction setting out key provisions and changes, it explains some basic legal concepts and key legislation relevant to smaller projects. As the form is short, it is particularly important to understand the legal framework within which it is set, as this will help to ensure it is used appropriately.

For example, MW16 complies with the Housing Grants, Construction and Regeneration Act 1996, as amended by the Local Democracy, Economic Development and Construction Act 2009. The provisions introduced by this legislation, which ensure the form can be used on commercial projects, are identified and discussed in the Guide. However, these provisions are not essential for contracts not covered by this Act, such as those with a residential occupier. Therefore, where the form is being considered for use in this situation the complex provisions and their implications should be explained carefully to the client in advance of its selection.

The Guide examines the key issues that would be relevant to the contract administrator and the parties to the contract: the programming and timing issues, quality and control of the works, payment, insurance, termination and dispute resolution. It contains many tables that might act as useful checklists for the practitioner and diagrams to clarify the procedural sequences.

This Guide is intended as a clear and straightforward point of reference for those using and studying the form, including those encountering the contract for the first time and those needing regular reminders when actively engaged in administering a project.

1 About MW16

1.1 There are two versions of the Minor Works Building Contract 2016, one that includes provision for the contractor to carry out design (the 'Minor Works Building Contract with contractor's design', MWD16) and one that does not (MW16). Apart from the provisions relating to design, the two versions are identical and can be used by private clients and public or local authorities alike. Therefore, although this guide refers to MW16 throughout, the points made are equally applicable to both versions. Where matters relating to MWD16 only are covered, these are clearly distinguished.

1.2 Separate editions of the Minor Works Building Contract are published by the Scottish Building Contract Committee (SBCC) for use in Scotland (at the time of writing the most recent versions available are the Minor Works Building Contract 2011 and the Minor Works Building Contract with contractor's design 2013, although it is very likely these will be updated to 2016 versions in the near future). The key differences in the Scottish versions are the absence of the attestation, the inclusion of provisions for bills of quantities, an additional Schedule for listing the contract documents and detailed guidance on 'signing' under Scottish law.

1.3 There are no separate supplements published for use with MW16. Provisions concerning arbitration and fluctuations are included as Schedules 1 and 2 at the back of the form, and Schedule 3 sets out eight optional supplemental provisions. The form also includes guidance notes. The JCT publishes a Minor Works Sub-Contract with sub-contractor's design (MWSub/D16) for use with MWD16 where the sub-contractor is undertaking design. The generic JCT Short Form of Sub-Contract (ShortSub16) would be suitable for use alongside MW16, as would the JCT Sub-subcontract (SubSub16). The JCT has helpfully published 'Tracked Change' versions of MW16 and MWD16 for those who wish to see where changes have been made from the 2011 editions. In addition, various free publications that may be very helpful are available to download from the JCT website, such as user checklists.

Key features

1.4 MW16 is a traditional lump sum contract, relatively simple in its overall structure and with few procedural rules. The contractor undertakes to carry out the work shown in the contract documents identified in the first recital, by the completion date entered in the contract particulars, in return for a contract sum entered in Article 2. There are provisions for varying the work, together with mechanisms for adjusting the contract sum and the completion date. In general terms, the contract assumes that all work is designed by the contract administrator. Under the MWD16 version there is provision whereby design responsibility can be assigned to the contractor for an identified part or parts of the works – termed the 'Contractor's Designed Portion' (the 'CDP'). Apart from this provision, the forms assume that the contractor will have no design role.

1.5 The form requires the appointment of an 'Architect/Contract Administrator', who is responsible for issuing all further information necessary for the carrying out of the works (for simplicity, this Guide refers only to 'contract administrator' throughout). The person's name and firm are entered in article 3, and footnote 7 indicates that this should not be the employer, unless the contractor has agreed to this appointment. At some points the contract administrator is acting as the employer's agent, and at others as an independent administrator. A court would assume that the parties have contracted on the basis that the contract administrator will act fairly at all times in applying the terms of the contract, and particularly so when deciding such matters as payments due to the contractor and extensions of time. Any failure in this duty would amount to a breach of contract by the employer, which may result in a claim by the contractor against the employer and, should the contractor be successful, a claim by the employer against the contract administrator. This duty of fairness, does not place the contract administrator in the same position as an arbitrator, in that the contract administrator is not immune from being sued.

1.6 The contractor takes full responsibility for ensuring that the standards set out in the contract documents are achieved, and this includes direct responsibility for any sub-contracted work. The form provides for domestic sub-contractors chosen by the contractor subject to the written consent of the contract administrator. There is no provision by which the contractor can be required to use a sub-contractor named or nominated by the contract administrator. The guidance notes to the form indicate that a firm could be named in the tender documents, or in an instruction in relation to a provisional sum, but also explain that there are no terms that deal with the consequences of such naming. While it may be possible to do this, there will always be some uncertainty as to the outcome, and so risk to the employer. Therefore, if the ability to name a sub-contractor is crucial it may be better to use another form (see Table 1.5).

1.7 The work to be undertaken will be in accordance with the contract documents, but will also include varied or additional work subsequently instructed as provided for in the conditions. The contract sum may be adjusted and is qualified by the wording 'or such other sum as becomes payable under this Contract' (Article 2). The amount of work which is covered by the contract sum should be described in exact terms in the contract documents. The work is described in drawings and/or a specification and/or schedules of work and, where the CDP is used, the employer must state its requirements for the design of that part (the 'Employer's Requirements'). If a BIM protocol is to be used, the JCT guidance notes recommend that this is included in the employer's requirements (for MWD16) or in the contract documents (for MW16). The contractor prices the specification or work schedules or provides a schedule of rates. Generally, if the description is inaccurate, any resulting addition to the cost is borne by the employer. If the contractor has made an error in pricing, however, then any shortfall will be borne by the contractor. The contract administrator has wide powers to order variations to the works if required, and the contractor has a corresponding right to be paid any additional costs that arise from such variations.

1.8 Interim payments are made to the contractor at monthly intervals (unless agreed otherwise) following the issue of contract administrator's certificates. In general terms, the certificates will reflect the amount of work that has been properly completed up to the point of valuation in accordance with the terms of the contract. None of these certificates are conclusive evidence that the contractor has fulfilled its obligations under the contract. There is no reference to the appointment of a quantity surveyor, and there are no functions assigned to such a person in the conditions. The contract administrator therefore has a significant role, which extends to the valuation of variations, the valuation of work executed and the computation of the final sum due. The contract administrator and contractor are required to 'endeavour to agree' on the price of a variation before the work is carried out; if no agreement is reached these are valued by the contract administrator. If a quantity surveyor is appointed by the employer, the quantity surveyor would act solely as adviser to the contract administrator and employer and would have no active role under the form.

1.9 The form complies with the Housing Grants, Construction and Regeneration Act (HGCRA) 1996 (as amended by the Local Democracy, Economic Development and Construction Act 2009). It also includes provisions aimed at ensuring it is appropriate for use by local authorities. These are discussed below at paragraphs 1.39 and 1.19 respectively.

Changes since MW11

1.10 Prior to the 2016 editions the JCT had published one set of adjustments to the 2011 editions: Amendment 1 (March 20015), relating to the CDM Regulations 2015. The JCT had also published a Public Sector Supplement for use by local and public authorities. These changes are now incorporated in the 2016 editions.

1.11 The key 2016 changes are set out in Table 1.1 and can be summarised as follows:

- incorporation and updating of provisions from the JCT Public Sector Supplement relating to Fair Payment principles and transparency;
- amendments relating to the CDM Regulations 2015;
- reference made to various provisions of the Public Contracts Regulations 2015;

Table 1.1 Key changes

Clause	Revised/new	Key changes
Agreement	revised	Re-titled: previously 'Articles of Agreement'
Article 4	revised	'CDM Co-ordinator' replaced with 'Principal Designer' (this change is made throughout the form)
Contract particulars	revised	Seventh recital (eighth recital in MWD16): two new supplemental provisions added Cl 4.3: 'Interim Valuation Dates' added Cl 5.4C: provision for entering the parties' insurance arrangements
1.1	revised	New definitions of: Contractor's Persons, Employer's Persons, Interim Valuation Date, Local or Public Authority, PC Regulations, Statutory Undertaker, Works Insurance Policy
1.7	new	Consents and approvals shall not be unreasonably delayed or withheld
2.1	revised	Redrafted but no significant change; footnote on BIM added to MWD16
2.10 (MWD16)	revised	Practical completion: requirement for supply of contractor's design drawings, etc. added
3.3.1 (MWD16)	revised	Requirement for consent to sub-contract design work added
3.6.3	revised	Valuation of variations: 'any matters that are to be treated as a variation' added

Table 1.1 Key changes – Continued

Clause	Revised/new	Key changes
3.9	revised	Changes required due to CDM Regulations 2015; also drafting simplified
4.3	revised	Certification and payment related to 'Valuation Dates'; interim certificates continue during the rectification period; reference to clause 4.7 (costs/expenses caused by suspension) added
4.4 (MW11)	deleted	Clause relating to 'penultimate certificate' deleted
4.4	new	Contractor's right to make payment application
4.5 and 4.6	revised	Redrafted but no significant changes
4.8	revised	Drafting simplified as clauses 4.5 and 4.6 (notices and failure to pay apply also to final payments)
4.9 (MW11)	deleted	Clause relating to failure to pay final amount deleted. Clause 4.6 (failure to pay amount due) applies also to final payments
5.2	revised	Contractor's liability for damage to existing structures clarified
5.4C	revised	New text referring to parties' agreed insurance policy requirements
5.6	new	Consolidated clause dealing with action following damage to works under all insurance options
5.7	new	Right to terminate following material loss or damage to existing structures
6.1	revised	New definition of insolvency
6.6 and 6.10	revised	Reference to Public Contract Regulations 2015 ('PC Regulations') added
Schedule 3	revised	New supplemental provisions relating to transparency (Freedom of Information Act 2000) and the Public Contract Regulations 2015 added

- changes in respect of payment, designed to reflect Fair Payment principles and to simplify and consolidate the payment provisions (e.g. interim certificates continue monthly throughout the rectification period); and
- clause 5.4C (works insurance) revised to allow parties to specify their own insurance policy requirements.

Use in different situations

1.12 MW16 is intended for projects that are procured on what is normally referred to as a 'traditional' procurement route, i.e. one where the client engages at least one, and possibly several, firms of consultants to prepare a design and complete full technical documentation before the project is tendered to contractors. Traditional procurement is widely used, especially on smaller projects (it was applied in around 76 per cent of projects undertaken in the UK in 2010, which accounted for 41 per cent of the total value: RICS and Davis Langdon, 2012). A headnote at the front of the form states that it is appropriate for work that is simple in character, where the work is designed by or on behalf of the employer and where a contract administrator has been appointed to administer the conditions.

1.13 The guidance notes included at the back of the form add that it is intended that a lump sum offer will have been obtained at tender stage, based on documents sufficiently detailed for the contractor to be able to price accurately without a bill of quantities. The anticipated contract period should be such that full fluctuations provisions are not required (in practice it would be appropriate for programmes of up to around 12 months' duration). JCT Practice Note 5 (series 2) stated that MW98 was 'suitable for contracts up to the value of £100,000 (2001 prices)', but the subsequent versions of this Note – Deciding on the Appropriate JCT Contract – have not set out any limits. It is generally recognised that it is the nature of the project, rather than its value, that should be the deciding factor, and that the form can be used successfully on larger projects provided they are straightforward.

1.14 The headnote makes it clear that the form is not suitable for use as a design and build contract. In such cases, the JCT Design and Build Contract or the JCT Constructing Excellence Contract would be the appropriate choice for larger projects. For smaller projects, it may be possible to adapt one of the RIBA Building Contracts.

1.15 In the case of MWD16, however, the headnote adds that this version is suitable for projects where the contractor is required to design discrete part(s) and the employer has had detailed requirements prepared for that design. It is common that some parts of the design are completed by the contractor or by specialist sub-contractors (this is sometimes referred to as 'traditional plus contractor design'), but if this is to take place, a suitable form must be used: MWD16 would be suitable where the contractor's design input is limited to one or two simple and non-critical elements, as the provisions relating to contractor design are quite brief. For example, there is no provision for requiring professional indemnity insurance from the contractor. The procedures for the contractor to submit the developing design are limited – the interval between submission of information and starting the relevant work may be as little as seven days – and there is no system by which the contract administrator can comment or require changes. If the design elements are complex or critical, and/or such procedures are required, it may be better to use the JCT Intermediate Building Contract with contractor's design (ICD16), with its more detailed CDP provisions, or, for smaller projects, the RIBA Building Contracts.

1.16 As noted above, MW16 does not contain any clauses whereby the employer can require that the contractor uses a particular sub-contractor. If a firm is named in the tender documents, and it is made clear that the contractor will accept full responsibility, the risk to the employer is reduced, for example a delay caused by the firm should not give rise to an extension of time. However, if the sub-contractor is introduced to the project after tender, say by means of an instruction, the effects will be far less clear. It is important that careful thought is given as to the motives for such naming. Usually, the reason for involving a firm relates to a detailed design matter, where the consultant designers have sought design input from a specialist company. However, the form does not expressly cater for this type of arrangement, and does not provide for any design warranties. Should defects later emerge, the employer, if it cannot claim against the specialist firm, and if the contractor is arguing it is not liable, is very likely to hold the contract administrator responsible. Taken together, it may be better to avoid naming, or to use a form that includes full provisions.

1.17 MW16 is not intended to be used with management contracting or construction management arrangements (where the project is tendered as a series of packages to separate firms, with work progressing on a rolling programme basis after the first package is let). Management procurement arrangements are normally used on very large projects; however, on a smaller scale, clients who wish to manage projects themselves sometimes adopt a similar system and engage a number of separate companies (often referred to as 'separate trades'). MW16 might be suitable for some of the larger work packages, but thought should be given as to how all the separate contracts are to be co-ordinated. This is not an easy task, and the apparent savings achieved by cutting out the main contractor's markup may be more than offset by the amount of time the client has to spend managing the process, or by the additional fees charged by consultants if they undertake this role.

Use by domestic clients

1.18 MW16 may frequently be considered as an option for domestic building works. A key issue in such cases is whether the client is a 'residential occupier' for the purposes of the HGCRA 1996 or a 'consumer' for the purposes of the Consumer Rights Act 2015. If the client is a residential occupier, then the provisions regarding payment notices and adjudication are not required by law (see paragraphs 1.39–1.42), and some thought should be given as to whether the client would wish to include them or would prefer a simpler payment regime and an alternative dispute resolution system. Furthermore, if the client is a consumer (an individual acting outside their normal business, see

Table 1.2 Key duties of the employer

Clause (MW/MWD)	Employer's duty
2.8.3/2.9.3	Notify contractor of intention to deduct liquidated damages
3.9	Comply with the CDM Regulations, ensure the principal designer carries out his or her duties, notify the contractor of any replacement of principal designer or principal contractor
4.1	Pay the contractor VAT
4.5.1	Pay sums due as stated in certificates
4.5.2	Pay sums due as stated in a contractor's payment notice
4.5.4	Give the contractor a pay less notice, if intending to pay less than sum certified
4.6.1	Pay interest on amounts not paid
5.4B	Take out insurance against loss or damage to the works and existing structures as required
5.4C	Take out insurance as required in the contract particulars
5.5	Produce evidence of insurance
5.6.5	Pay all insurance monies to the contractor
6.11.4	Pay the contractor the amount properly due
Supplemental Provision 1	Work collaboratively with other team members
Supplemental Provision 2	Establish a working environment where health and safety is of paramount concern
Supplemental Provision 3	Confirm a cost-saving measure in an instruction
Supplemental Provision 4	Monitor and assess the contractor's performance
Supplemental Provision 6	Notify the contractor of any matter that may give rise to a dispute, meet and engage in good faith negotiations to resolve a dispute
Supplemental Provision 7	Inform the contractor of any request for disclosure

Table 1.3 Key powers of the employer

Clause (MW/MWD)	Employer's express power
2.8/2.9	Require the contractor to pay or allow the employer liquidated damages; deduct the damages from any sums due
3.5	Employ and pay others to execute work, if the contractor fails to comply with an instruction
5.7	Terminate the contractor's employment
6.4	Terminate the contractor's employment
6.5	Terminate the contractor's employment
6.6	Terminate the contractor's employment
6.7	Employ and pay others to complete the works
6.10	Terminate the contractor's employment

paragraph 1.45), it should be borne in mind that MW16 has not been drafted as a consumer contract. While it includes the HGCRA 1996 provisions, it makes no reference to the Consumer Contracts (Information, Cancellation and Additional Charges) Regulations 2013, which give the consumer the right to cancel the contract within 14 days of signing it. None of these issues mean that it cannot be used, but the provisions and their implications should be explained carefully to the consumer, and it should not be recommended or selected simply because this has been the usual practice of the client's advisers on smaller projects. The key duties and powers of the employer are set out in Tables 1.2 and 1.3.

Use by public bodies

1.19 As noted above, MW16 has many features required for public sector procurement (which are missing from other shorter forms) relating to Fair Payment provisions and transparency. Some were originally published as a supplement to the JCT contracts, including provisions for adoption into MW16, but have now been incorporated into the form with some further amendments. The 'Fair Payment' provisions arise from the stated aims of the government in *Construction 2025*, which include equitable financial arrangements and certainty of payment throughout the supply chain. The aims are reflected in initiatives such as the Construction Supply Chain Payment Charter 2014, the HGCRA 1996 (as amended), the Late Payment of Commercial Debts Regulations 2013 and the Fair Payment Charter, as well as regulation 113 of the Public Contracts Regulations 2015. These require that the final date for payment should not exceed 30 days from the date at which the value of work done in a particular period is assessed, and that similar provisions are included in sub-contracts and sub-subcontracts (MW16 Supplemental Provision 8). Under the charter, the value of work and materials supplied by all three tiers is to be assessed as at the same date. Adopting MW16 together with the appropriate JCT sub-contracts will ensure that these requirements are met. The contract also includes provisions (Supplemental Provision 7) relevant to any employer that is subject to the Freedom of Information Act 2000 (which would include local authorities). This provides that the parties accept that the contract is not confidential, except for material that may be 'exempt' and which the employer has the discretion to determine. The Public Contracts Regulations also deal with awarding of contracts, corrupt practices and bribery, and breach of the statutory requirements is a ground for termination under clauses 6.6 and 6.10.3 of MW16. Furthermore, under Supplemental Provision 8 the contractor must include similar provisions in any sub-contract.

1.20 MW16 is therefore an appropriate choice for smaller or more straightforward local authority projects, such as for stripping out or site clearance work in preparation for a larger project to be let on the JCT Standard Building Contract, or equivalent (reference to a framework agreement is included in the sixth recital (seventh recital in MWD16), which may be relevant to local authorities and public bodies). In some cases where projects have been let on a 'letter of intent', it may have been far better had a JCT Minor Works contract been used in the short term, while the parties finalised the terms for the main works.

Advising on MW16

1.21 One of the most striking characteristics of MW16 is its brevity, at least in comparison with most other standard forms. Its popularity suggests that this apparent advantage is appealing to many. However, the fact that the contract is silent on many matters can create problems. The procedural rules are minimal, and there is little precise information on the content, form and timing of notices, etc., leaving the parties to agree their own in advance or to sort matters out as they proceed. The contract is also silent on the rights of the parties should certain circumstances arise, many of which are quite common in practice. This will not mean that the parties have no rights, as in many cases the courts would imply a term into a contract to cover the particular situation.

1.22 Somewhat paradoxically, with MW16 there is a greater need to understand the general principles of contract law, in particular the law relating to implied terms, than there is with forms of contract that are longer and more sophisticated. Unfortunately, this fact is sometimes not appreciated, and there have been notorious cases of architects mishandling administration under the Agreement for Minor Building Works, often through failing to understand the legal framework in which it operates. This chapter includes a short summary of the principles of implied terms, which are referred to at several points in the Guide. However, if complex issues arise, the contract administrator should be prepared to take legal advice.

1.23 Before advising on whether or not to use MW16, the contract administrator needs to be as aware of what is missing from the form as of what is included, and to consider whether the absence of matters not expressly included might cause difficulties. For example: Is it important to use certain sub-contractors or suppliers?; Will there be a design element in some specialist sub-contract work which should be covered by a warranty?; Does the employer intend to remain in occupation throughout the work, thereby involving the contractor in some decanting and phased programming of the works? Consideration should also be given to whether some or all of the Schedule 3 Supplemental Provisions 1 to 6 should be included: some may be helpful for most projects (e.g. collaborative working) whereas others may be overly complex if the project is quite small (e.g. performance indicators). If the employer is a local or public authority, its attention should be draw to Supplemental Provisions 7 and 8.

1.24 It is often the case that minor building work will be commissioned by less experienced clients for 'once in a lifetime' operations. Although the amounts of money involved may be relatively small, to many employers they are a very significant expenditure. Despite this context, in some areas the contract is less protective of the employer than one might expect. For example, although the JCT Standard Building Contract (SBC16) requirements for a right to interest in sub-contracts (protecting the sub-contractor) has been stepped down into MW16, the sub-contract conditions designed to protect the employer in regard to ownership of unfixed materials have not. It is therefore important for the contract administrator to take time and care to explain to the client the respective obligations of the parties, particularly in respect of matters such as insurance and payment of sums properly due. While such matters as extensions of time and reimbursement of loss and expense are covered in the form, the procedures are somewhat slight. The employer may therefore be unaware at the time of any delays in progress, as MW16 does not require the contractor to notify the contract administrator until it becomes apparent that completion will not be achieved. Even then, it seems that this operates only when the delay is due to reasons 'beyond the control' of the contractor.

1.25 For larger or more complex projects, JCT Intermediate Building Contract (IC16) or ICD16, which contain all the provisions listed in Table 1.4, may therefore be a more appropriate

Table 1.4 Key provisions not included in MW16
• a clerk of works
• phased possession or completion
• use or occupation of the site by the employer during the works
• nominated or named sub-contractors or suppliers
• the contractor to notify the contract administrator of discrepancies between the contract documents
• the employer to employ others directly, or the contractor to allow access for directly engaged workers
• the contract administrator to reduce the contract period if work omitted
• the contract administrator to award an extension of time without a notice from the contractor (but see paragraph 4.23)
• the contract administrator to extend the contract period for delaying events occurring after the date for completion has passed
• the contract administrator to review extensions of time previously given
• the contract administrator to award loss and/or expense, unless resulting from a variation or suspension
• the contract administrator to require work to be opened up, tested or removed
• the contract administrator to visit the contractor's or sub-contractors' workshops
• the contractor to include 'property vesting' clauses in all sub-contracts
• professional indemnity insurance requirements for contractor's designed portion
• design submission procedure for contractor's designed portion
• copyright in the contractor's design documents
• provisions for collateral warranties from the contractor or any sub-contractor
• provisions for advance payment
• allowance for terrorism insurance cover
• a quantity surveyor

Table 1.5 Comparison of standard forms

	JCT Intermediate Building Contract with contractor's design 2011 (ICD11)	JCT Minor Works Contract with contractor's design 2016 (MWD16)	JCT Home Owner Contract	NEC3 Short Contract	RIBA Concise Building Contract 2014	RIBA Domestic Building Contract 2014	FMB Domestic Contract for Minor Building Work (2011)	FMB Domestic Building Contract (2011)
Progress meetings					✓			
Contractor design	✓	✓		✓	option	option	option	option
Employer-selected sub-contractors	✓				option	option	option	option
CDP submission	✓	✓			✓	✓		
Professional indemnity insurance	✓				✓	✓	✓	✓
Programme				option	option	option		
Sectional completion	✓				✓	✓		
Partial possession	✓				✓	✓		
Variation quotation				✓	✓	✓		
Loss/expense	✓	limited	limited	✓	✓	✓	✓	✓
Variation and extension of time rules					option	option		
Retention	✓	✓	✓	✓	✓	✓	✓	✓
Pay less notice	✓	✓		✓	✓		✓	✓
Interest	✓	✓		✓	✓		✓	✓
Customer cancellation			✓			✓	✓	✓
Fair dealing clause	option	option						
Key performance indicators	option	option						
Negotiation	option	option						
Public sector clauses	✓	✓						

choice (see also Table 1.5). For smaller projects, particularly those for domestic work, the clear and simple Home Owner Contracts (HOC) may be used. These do not contain the HGCRA 1996 provisions (as amended), as discussed below (as they are not required on projects where the employer is a residential occupier), but they do have some significant limitations, such as the lack of provisions for liquidated damages. For very small commercial repair jobs, the Repair and Maintenance Contract (Commercial) (RM) could be used, with one version for domestic and one for commercial work. The RIBA Building Contracts for domestic projects and for commercial work are also viable alternatives which are proving popular in practice, but of course to date they have only a short track record. The fact that the JCT MW forms have been around since 1968 means that the meaning of many clauses has been tested and clarified in the courts and, where problems have been experienced in practice, the JCT has amended and fine-tuned the provisions as necessary.

1.26 In many cases, therefore, MW16 or MWD16 will be an entirely appropriate selection, and the form justifies its popularity. It is simple to complete, easy to refer to and generally much less daunting to the less experienced client or builder than many other building contract forms. The drafting in the recent editions has been clarified and, in places, shortened significantly. It is of course possible to introduce amendments to tailor the form to particular situations, but this should not be done without legal advice. If major changes are needed then a more appropriate form of contract should be considered. Otherwise, the contract administrator should not be concerned about recommending MW16 for the type of straightforward project for which it was intended.

Some general principles of contract law

Formation

1.27 A contract is formed when an unconditional offer is unconditionally accepted. In the context of a building project, where contractors have been invited to submit competitive tenders, the tenders constitute an 'offer' to carry out the work shown in the tender documents for the price tendered. If a tender is accepted then a contract will have been formed, and the terms of the contract will be those set out or referred to in the tender documents.

1.28 'Letters of intent' can cloud the picture and should be avoided. If it is possible to accept the tender without qualification then it is better simply to write a letter to that effect, and the contract comes into existence from the moment the letter has been received by the contractor. The effect of a letter expressing an intention to enter into a contract at some point in the future will depend on the wording and circumstances in each case, but it is likely to be of no legal effect. Starting work on such a basis could have disastrous consequences for both parties.

1.29 If there is a period of negotiation, careful records should be kept of all matters agreed in order that they can be accurately incorporated into the formal contract documents. These documents should always be prepared as soon as agreement is reached, and before work commences on site. Failure to execute the documents does not necessarily mean that no contract is in existence, but it might give rise to sufficient doubt to require spending on legal fees in an effort to establish the true position (*Goldsworthy* v *Harrison*). In addition, it can often lead to avoidable arguments about what was agreed. The contract, once executed, will supersede any conflicting provisions in the accepted tender and will apply retrospectively (*Tameside Metropolitan BC* v *Barlow Securities*).

Goldsworthy and others v *Harrison and another* [2016] EWHC 1589 (TCC)

The case concerned the enforcement of an adjudicator's decision. The defendant employers were residential occupiers and the key matter in dispute was whether the parties had agreed contract terms that contained an adjudication clause. If they had not, the adjudicator had no jurisdiction. At the time of tender, there had been no mention of use of JCT MW. However, an email around the time work began stated 'As discussed previously the contract will be a JCT Minor Work', which was followed shortly after by another stating 'If you are successful in the quotation for the garage and summer house, and main works please note at that time there will be retentions applied and the JCT Minor works contract'. However the documents were never executed, and there was no evidence that key matters such as liquidated damages had been agreed. The court refused to give summary judgment, stating that without fuller evidence from both sides, in particular of the discussions lying behind the emails, it was impossible to say that there was not a triable issue on the question of whether the parties had agreed on the JCT Minor Works terms, in particular the gaps where particular options had not been filled in or agreed. The matter would therefore have to proceed to a full trial.

Tameside Metropolitan Borough Council v *Barlow Securities Group Services Limited* [2001] BLR 113

Under JCT63 Local Authorities, Barlow Securities was contracted to build 106 houses for Tameside. A revised tender was submitted in September 1982 and work started in October 1982. By the time the contract was executed, 80 per cent of building work had been completed, and two certificates of practical completion were issued relating to seven of the houses in December 1983 and January 1994. Practical completion of the last houses was certified in October 1984. The retention was released under an interim certificate in October 1987. Barlow Securities did not submit any final account, although at a meeting in 1988 the final account was discussed. Defects appeared in 1995, and Tameside issued a writ on 9 February 1996. It was agreed between the parties that a binding agreement had been reached before work had started, and the only difference between the agreement and the executed contract was that the contract was under seal. It was found that there was no clear and unequivocal representation by Tameside that it would not rely on its rights in respect of defects. Time began to run in respect of the defects from the dates of practical completion; the first seven houses were therefore time barred. Tameside was not prevented from bringing the claim by failure to issue a final certificate.

Express terms

1.30 Generally speaking, parties are bound by the terms of the contract which they have expressly set out and agreed. In practice there may be difficulties in establishing exactly what these terms are: they may be scattered among several documents, they may be ambiguous or contradictory or they may be silent on some aspect of the matter under dispute. The process of piecing together and interpreting the terms of a contract is governed by a distinct area of law. Some of the more important rules are as follows:

- Words should be given their ordinary literal meaning; where there is ambiguity or a conflict, generally a court will determine, on an objective basis, what it considers were the true intentions of the parties. For example, specially agreed terms will normally prevail over standard printed terms, as these are more likely to represent the parties' intentions. However, this would not apply to MW16 as it contains a clause which states that the printed terms prevail.
- The contract is usually construed most strongly against the party which drew it up (termed the *contra proferentem* rule). It is generally considered, however, that this rule would not apply to JCT standard forms that are negotiated, rather than drawn up by one party. It may nevertheless apply to other contract documents, such as the specification, unless the terms in question have been specifically negotiated, and may also apply to the form itself if it has been amended in significant respects.

- Generally speaking, evidence of previous negotiations is not admissible to contradict the express terms of the contract (*Wates Construction* v *Bredero Fleet*), though evidence of the factual background may be used in relation to implied terms (see paragraph 1.31).

Wates Construction (South) Ltd v *Bredero Fleet Ltd* (1993) 63 BLR 128

Wates Construction entered into a contract on JCT80 to build a shopping centre for Bredero. Some sub-structural work differed from that shown on the drawings and disputes arose regarding the valuation of the works, which were taken to arbitration. In establishing conditions under which, according to the contract, it had been assumed the work would be carried out, the arbitrator took into account pre-tender negotiations and the actual knowledge that Wates gained as a result of the negotiations, including proposals that had been put forward at that time. Wates appealed and the court found that the arbitrator had erred by taking this extrinsic information into consideration. The conditions under which the works had to be executed had to be derived from the express provisions of the bills, drawings and other contract documents.

Implied terms

1.31 In addition to the interpretative rules outlined above, there are several mechanisms whereby terms which the parties have not expressly set out may be implied into a contract.

1.32 A term can be implied 'in fact' or 'in law'. Terms are implied in fact to give effect to the presumed but unexpressed intentions of the parties and will not be implied if they would contradict the express terms. They are implied on the basis of the particular circumstances of that contract and normally must survive a 'test of necessity'; in other words, that without the implication the contract would be so unbusinesslike that no sensible person would ever have agreed to it. The courts have not always applied the test with this degree of stringency, and will sometimes imply a term on the basis that it appears the parties intended it.

1.33 In addition, the courts' approach to the range of circumstances that can be looked at, sometimes referred to as the 'factual matrix', has varied considerably from a broad approach taking into account a wide variety of surrounding circumstances, to a very narrow one which confines itself to the 'four corners' of the contract documents. (A typical 'narrow' approach is shown in *Wates Construction* v *Bredero Fleet* in paragraph 1.30.) In practice, it would be unwise to rely on a term being implied on the basis of the surrounding circumstances.

1.34 Terms are implied in law where either (a) they are always implied into that type of contract as a matter of legal incidence or (b) through the operation of statute. In either case the term is not based on the presumed intention of the parties. The fact that a term contradicts the express terms of a contract will not necessarily prevent its being implied. As an example of terms implied as a necessary incidence, certain obligations would always be implied into contracts between landlord and tenant. By far the most important implied terms with respect to construction contracts are those implied by statutes. The most significant of these statutes are the Sale of Goods Act 1979, the Supply of Goods and Services Act 1982 (both amended by the Sale and Supply of Goods Act 1994), the Defective Premises Act 1972, the HGCRA 1996 (as amended) and the Consumer Rights Act 2015.

1.35 Conversely, statute also operates to curtail what may be included in contracts, i.e. it limits in certain circumstances the introduction of provisions that might be considered against public policy, or simply unfair. The primary means of achieving this is through the Unfair Contract Terms Act 1977 (for commercial contracts) and the Consumer Rights Act 2015 (for consumer contracts). It is important to be aware of what may be implied and what may be considered unfair as this may affect which contract form should be selected for use in a particular situation, and it is a critical matter to be considered if amending a standard form.

The Sale of Goods Act 1979

1.36 This implies terms into contracts for the sale of goods regarding title (section 12), correspondence with description (section 13), quality and fitness for purpose (section 14) and sale by sample (section 15). For example, section 14 implies a term that where the seller sells goods in the course of business and the buyer, expressly or by implication, makes known to the seller any particular purpose for which the goods are being bought, there is an implied condition that the goods supplied under the contract will be reasonably fit for that purpose.

The Supply of Goods and Services Act 1982

1.37 This covers contracts for work and materials, contracts for the hire of goods and contracts for services. Most construction contracts come under the category of 'work and materials' and the Act implies terms into these equivalent to sections 12 to 15 listed in paragraph 1.36 with respect to any goods in which the property has been transferred under the contract. So, as above, any materials supplied should be reasonably fit for their intended purpose, provided always that the buyer is relying on the supplier's skill and judgment. (If the buyer specifies a particular material then this would be sufficient to show that it was not relying on the seller.) For services, the Act implies terms regarding care and skill, time of performance and consideration. For example, section 14 implies a term that where the supplier is acting in the course of business, the supplier will carry out the services within a reasonable time, provided of course the parties have not themselves agreed terms regarding time.

The Defective Premises Act 1972

1.38 This applies where work is carried out in connection with a dwelling, including design work. It states that 'A person taking on work for or in connection with the provision of a dwelling … owes a duty … to see that the work which he takes on is done in a workmanlike or, as the case may be, professional manner, with proper materials and so that as regards that work the dwelling will be fit for habitation when completed' (section 1(1)). This appears to be a strict liability, and is owed to anyone acquiring an interest in the dwelling.

The Housing Grants, Construction and Regeneration Act 1996

1.39 The HGCRA 1996 (as amended by the Local Democracy, Economic Development and Construction Act 2009) applies to most construction contracts (the term 'construction contract' is given a wide definition, and includes contracts for services only or for building work, and applies to a large range of types of work, including demolition works, services installations and repair work, as well as new buildings, extensions and alterations). The Act requires that construction contracts include specific terms relating to adjudication and payment:

- the right to stage payments;
- the right to notice of the amount to be paid;
- the right to suspend work for non-payment;
- the right to take any dispute arising out of the contract to adjudication.

1.40 If the parties fail to include these provisions in their contract, the Act will imply terms to provide these rights (section 114) by means of the Scheme for Construction Contracts (England and Wales) Regulations 1998.

1.41 However, there is an important exception; it does not apply to projects where one of the parties is a 'residential occupier'. The residential occupier exception applies to projects for which the primary purpose is beneficial use by the employer as a residence (section 106). This would include buildings that the client is occupying or intending to occupy as its main residence, and might also include a second home if the client is the main user and there is no intention to use it as a holiday let (*Westfields* v *Lewis*). However, work to buildings in the grounds of a residence that will not be lived in by the customer, or work to divide a property into flats where only one flat will be retained by the customer, will not fall under the exception (*Samuel Thomas* v *Anon*). Similarly, work on other residential properties, for example for landlords, local authorities or housing associations, will usually be covered by the Act.

Westfields Construction Ltd v *Lewis* [2013] BLR 233 (TCC)

This case related to the enforcement of an adjudicator decision. A key question was whether 'residential occupation' should be assessed at a single point in time (e.g. at the time the contract is formed) or as an ongoing process. The judge was of the view that 'occupies' is an ongoing process and must carry with it some reflection of the future: it indicates that the employer occupies and will remain at (or intends to return to) the property. The employer had moved out of the property and had talked about letting it. The judge found on the evidence that the employer intended to let out the property, and the contract therefore did not fall within the 'residential occupier' exception.

Samuel Thomas Construction v *Anon* (unreported) 28 January 2000 TCC

This case also related to the enforcement of an adjudicator decision. The contract was not on a standard form, and the judge had to decide whether the section 106 'residential occupier' exclusion applied. If it did the decision against the employer would be unenforceable. The contract concerned a number of buildings that were being refurbished, including a barn that the employer intended to occupy, and another barn and associated buildings that were being refurbished for sale. There was only one contract for the works. The judge upheld the adjudicator's view that where one dwelling was to be occupied and the other was not, the contract did not 'principally relate to operations on a dwelling which one of the parties ... intended to occupy' and therefore the exception did not apply.

Table 1.6 Provisions required under the HGCRA 1996 (as amended)

HGCRA 1996: section	MW16: clause	Provision concerning
108	7.2	Adjudication
109	4.3	Stage payments
110/110A	4.3/4.4/4.8	Dates for payment
111	4.5	Notices
112	4.7	Contractor's right of suspension
115	1.6	Notices

1.42 MW16 contains provisions that comply with the HGCRA 1996 (as amended) and it can therefore be used on any project, including those to which the Act applies (see Table 1.6). However, as these provisions introduce considerable complexity, it may be unwise to use the form on a project with a residential occupier, who may find the provisions onerous. As well as being difficult to operate in practice, depending on the circumstances some of the terms may also be caught by the unfair terms provisions in the Consumer Rights Act 2015 (see paragraph 1.48).

Exemption clauses

1.43 The scope for excluding liability for important matters is limited by two significant pieces of legislation, the Unfair Contract Terms Act 1977 and the Consumer Rights Act 2015.

Unfair Contract Terms Act 1977

1.44 This has the effect of rendering various exclusion clauses void, including any clauses excluding liability for death or personal injury resulting from negligence, any clauses attempting to exclude liability for Sale of Goods Act 1979 section 12 obligations (and the equivalent under the Supply of Goods and Services Act 1982), and any clauses attempting to exclude liability for Sale of Goods Act 1979 section 13, 14 or 15 obligations (and the equivalent under the Supply of Goods and Services Act) where they are operating against any person dealing as a consumer. It also renders certain other exclusion clauses void in so far as they fail to satisfy a test of reasonableness; for example, liability for negligence other than liability for death or personal injury, and liability for breach of section 13, 14 or 15 obligations in contracts which do not involve a consumer.

The Consumer Rights Act 2015

1.45 This Act, which came into force on 1 October 2015, consolidates much of the pre-existing legislation on consumer protection, and introduces some significant new provisions. It replaced the sections of the Unfair Contract Terms Act 1977 that relate to consumers, and repealed the Unfair Terms in Consumer Contracts Regulations 1999. It applies to contracts and notices between a 'trader' and a 'consumer'. A 'consumer' is defined as 'an individual acting for purposes that are wholly or mainly outside that individual's trade, business, craft or profession' (section 2(3)).

1.46 The Act applies to a wider range of contracts than other legislation commonly encountered by construction professionals. For example, under the HGCRA 1996 (as amended) only contracts with a residential occupier are excluded. It is therefore quite possible that the Consumer Rights Act will apply to a contract that is excluded from the HGCRA 1996 (e.g. where an individual undertakes work to a domestic property that is not the individual's main residence).

1.47 The Act states that any contract for services is to be treated as including a term that the trader must perform the service with reasonable care and skill (section 49(1)). In addition, if the contract does not provided for a price or timescale, it is taken to include a term that the services will be provided for a reasonable price (section 51), and within a reasonable timescale (section 52). Goods supplied under such a contract must also be of good quality.

1.48 Part 2 sets out the law regarding unfair terms in relation to consumers. Section 62(1) states that an unfair term of a consumer contract is not binding on the consumer. The test for 'unfair terms' in the Act is the same as that in the 1977 Unfair Contract Terms Act: it provides that a 'term is unfair if, contrary to the requirement of good faith, it causes a significant imbalance in the parties' rights and obligations under the contract to the detriment of the consumer' (section 61(4)). An 'indicative and non-exhaustive list' of examples of what might be considered unfair terms is set out in Schedule 2 to the Act. This includes, for example, any term which has the object or effect of excluding or hindering the consumer's right to take legal action or exercise any other legal remedy, which would include an arbitration agreement.

1.49 The most significant change relates to terms specifying the main subject matter of the contract or setting the price. These terms are *not* subject to the 'fairness' test provided that they are both transparent and prominent (section 64(1) and (2)). 'Transparent' is defined as being 'in plain and intelligible language and (in the case of a written term) legible' (section 64(3)) and 'prominent' as 'brought to the consumer's attention in such a way that an average consumer [who is reasonably well-informed, observant and circumspect] would be aware of the term' (section 64(4) and (5)). The new Act no longer makes an exception for terms that have been 'individually negotiated' as had been the case in the Unfair Terms in Consumer Contracts Regulations 1999.

1.50 Although the JCT does not state that MW16 has been drafted with the consumer in mind, the key agreements regarding subject matter and price are reasonably clear in the articles and contract particulars. The courts have, for example, held that consumer clients are bound by an adjudication agreement in a standard form contract, as it does not significantly affect the balance of power between the parties (*Lovell Projects* v *Legg & Carver*). It is possible, though, that some of the contract terms may be considered unfair (*Domsalla* v *Dyason* is a rare example of a case where this occurred). It would be sensible to go through the contract carefully with any consumer client and, if they are concerned by any of the provisions, to consider using another contract.

Lovell Projects v *Legg & Carver* [2003] BLR 452

The employers were residential occupiers and consumers who engaged a contractor to refurbish their house using JCT MW98. This edition included the JCT's own adjudication procedure (instead of relying on the Scheme for Construction Contracts, as it now does). The employers were advised by an architect, and during the tender negotiations they insisted that MW98 was used. Following an adjudication, the employers resisted enforcement of the decision on the basis that the adjudication provisions were unfair and had not been brought to their attention before they signed the contract. The court decided that contractual adjudication provisions in the consumer's contract were not unfair under the Unfair Terms in Consumer Contracts Regulations 1999 because they did not cause a significant imbalance in the parties' rights and obligations. The court bore in mind that it was the employers, not the contractor, who had proposed the term, and that the employers had access to advice from their contract administrator.

Domsalla v *Dyason* [2007] BLR 348

The employer, Mr Dyason, whose house had burnt down, was advised by his insurers to enter into a contract based on MW98 with the contractor, Domsalla. The employer saw the form for the first time at the first site meeting, when he was presented with it for signature. The project was delayed, and the employer did not pay the sums certified in the last three certificates for payment, totalling £127,871.33, but no withholding notices were served. The contractor initiated an adjudication and the adjudicator decided in its favour against the owner. Subsequently, the Technology and Construction Court refused to enforce the decision and the employer was given leave to defend the claim on the ground that the withholding notice provisions of the contract were unfair under the Unfair Terms in Consumer Contracts Regulations 1999, and were therefore not binding on him as a consumer.

2 Documents

2.1 Documents are central to the success of all building operations, even to small projects such as those carried out under MW16. In fact, with many small jobs on a short programme and tight budget, full and accurate information is essential right from the outset.

Tendering

2.2 It cannot be overemphasised that, when a project is sent out to tender, the contractor should be given full and detailed information regarding the project requirements. This is the principal means by which a designer can ensure that the quality of detailing, finish and workmanship will reach the required standards. If the project is not to be fully designed by the employer's consultants, the contractor will require full information about the design that it is to provide, including any performance specifications.

2.3 Most importantly, full details should be given about which standard building contract is to be used, the particular conditions to be applied and any special terms. Consultants tend to focus on the design, technical and budget aspects, but clarity on the terms of the contract is just as important to the tenderer as these will affect the tender price. It is not advisable to introduce matters such as special contract provisions after a long tendering period or after negotiations on price have been concluded: if they are not acceptable to the contractor, the contractor will be in a very strong bargaining position. Furthermore, if contract documents are not formally executed, the details included in the tender package may subsequently form the basis of the contract between the parties (see paragraphs 1.29 and 2.19).

Tendering procedures

2.4 Normally, for projects of the scale for which MW16 is intended, one of two methods will be used: competitive tendering with a small number of contractors, or negotiation with a single contractor. The employer's consultants will normally suggest which contractors should be considered, but the employer may have worked successfully with firms before, or may have been recommended a firm by others, in which case those firms can also be included.

2.5 With competitive tendering, all tenderers should, of course, be sent identical information, so that they are competing on an equal basis. The tenders are then examined by the employer's consultants, and normally the lowest priced one is accepted. With negotiated tendering, only the identified contractor will submit a tender, which is usually subject to discussion before acceptance.

2.6 An alternative method is a two-stage process: initially, a small number of contractors are asked to tender on the basis of less than complete information. Negotiations will then take place with the one that makes the most attractive submission. This method is frequently used where the contractor is expected to have a significant design input but the employer wishes to approve this design before entering into the main contract; the selected contractor can then work with the employer's consultants in finalising the design. It is unlikely that a two-stage tender would be used for a project based on an MW16 contract, but, if it is, an agreement will be needed to cover the contractor's liability to the employer for its contribution to the design process.

2.7 Full guidance on tendering procedures can be found in the *NBS Guide to Tendering: For Construction Projects* (Finch, 2011), JCT's *Tendering Practice Note 2012* (JCT, 2012) and the *RIBA Job Book* (Ostime, 2013). As noted above, the tender package should give full details of what will be included in the final contract package. The contract documents are outlined below, and Table 2.1 lists information that will be inserted in the contract particulars and indicates where it is discussed in this Guide.

Contract documents

2.8 MW16 defines 'Contract Documents' under the second recital (or the third in MWD16). This indicates that any or all of the following documents can be prepared on behalf of the employer:

- drawings;
- specification;
- work schedules;
- employer's requirements (MWD16 only).

2.9 The parties are required to delete the items that will not be used, and the remaining items, together with the agreement, the conditions and a schedule of rates (if used, see below) comprise the 'Contract Documents'. Interestingly, as 'drawings' are indicated as an item that may be deleted, it would in theory be possible to let the contract on a works schedule or on employer's requirements alone; in practice, however, drawings will almost always form part of the contract documents. Also, as the contract is not intended to be used as a 'design and build' contract, the employer's requirements will normally cover only part of the works, with the rest described by drawings plus schedules and/or a specification.

2.10 There is no provision for a bill of quantities. The contractor will normally price either the specification or the schedules in an itemised format, or provide a contract sum backed by a schedule of rates (third recital, or fourth recital in MWD16). Although in theory this recital could be deleted, and the contractor tender a lump sum figure on the basis of drawings alone, this would be impracticable as the priced document will be used, if relevant, in the valuing of variations. To be useful, therefore, the price breakdown will have to be reasonably comprehensive and detailed.

2.11 The 'Contract Drawings' are listed under the second recital (the third recital in MWD16). These should all be identified precisely, including revision numbers, etc. The list may be annexed if long, but if so, the list must be clearly identified. Note that in MW16 there is no reference to the party responsible for preparing the drawings.

2.12 A specification should normally form part of the documentation, as it is inadvisable to rely on notes on drawings except on the smallest of jobs. The Construction Project Information Committee (CPIC, now taken over by the NBS UNICLASS system) recommendation was that the specification becomes the core document in terms of defining quality, and that the drawings and other documents cross-refer to clauses in the specification. This is sensible advice and should prevent different standards being set in different documents.

2.13 MW16 does not describe what is meant by the term 'schedule', nor indicate what form the schedule should take. In practice it is generally a 'schedule of work', and could be arranged by work sections, by trades, or on a room-by-room basis as is common in refurbishment work. In all cases it is likely to be clearer if the detailed specification information is kept in a separate part of the document, or bound separately, and referred to by the schedules. Where two documents are used, it might be more satisfactory for the contractor to price the itemised schedule.

2.14 The contractor should be sent copies of all the intended contract documents at the time of tender, together with all the information required under the contract particulars, including commencement and completion dates (cl 2.2, or 2.3 in MWD16), liquidated damages (cl 2.8, or 2.9 in MWD16), the rectification period (cl 2.10, or 2.11 in MWD16), the retention percentage (cl 4.3), the period for supply of documentation (cl 4.8.1), together with information relating to tax, fluctuations, insurance and dispute resolution. For good practice in preparation and co-ordination of the specification, drawings and schedules, see current relevant CPIC publications on co-ordinated project information. If a BIM protocol is to be used, this should be included with the tender documents.

2.15 Table 2.1 lists information that will be inserted in the contract particulars.

Table 2.1 Information to be inserted in the contract particulars

Contract particulars entry (MW/MWD)	Details	Paragraph reference in this guide
Fourth/fifth recital and Schedule 2	Base date: the base date is usually set at around the time of return of tenders	6.17
Fourth/fifth recital and clause 4.2	Construction Industry Scheme: indicate if the employer is a 'contractor' for the purposes of the CIS	–
Fifth/sixth recital	CDM Regulations: indicate if project is notifiable. Note that the majority of CDM Regulations apply even when the project is not notifiable	3.20
Sixth/seventh recital	Framework agreement: insert details or refer to an identified and appended document	–
Seventh/eighth recital and Schedule 3	Supplemental provisions: Supplemental Provisions 1 to 6 apply unless indicated otherwise. Employers should consider whether these would be helpful and relevant to their project. (Supplemental Provisions 7 and 8 will apply where the employer is a public or local authority.)	1.19, 1.23, 4.14, 6.2, 10.4
Article 7	Arbitration: if arbitration is preferred to litigation as the dispute resolution method, select 'Article 7 and Schedule 1 apply'	10.3, 10.19

Table 2.1 Information to be inserted in the contract particulars – Continued

Contract particulars entry (MW/MWD)	Details	Paragraph reference in this guide
Cl 2.2/2.3	Works commencement date: the date from which the works *may* be commenced	2.14, 4.1–4.2
	Date for completion: the date by which the works *must* be completed (may be adjusted by an extension of time)	2.14, 4.1–4.2, 4.14
Cl 2.8/2.9	Liquidated damages: the rate is usually per day or per week, and should be an amount not 'out of all proportion' to the likely losses	2.14, 4.44
Cl 2.10/2.11	Rectification period: the default period is three months, but this is often not enough for anything but the smallest projects	4.38, 5.32
Cl 4.3	First interim valuation date: the default is one month after the works commencement date (note that the date remains fixed regardless of when the contractor actually starts work)	7.3
	Percentage of value of work included in interim certificates prior to practical completion: the default is 95 per cent. The total sum retained may amount to very little on a small project	2.14, 7.13
	Percentage of value of work included in interim certificates on/after practical completion	7.26
Cl 4.3 and 4.8	Fluctuations provisions: the parties may choose to have no fluctuations, select the Schedule 2 provisions (the default) or insert their own alternative requirements	6.16–6.18
Cl 4.8.1	Period for supply of documentation: normally the same as the rectification period, although a shorter period could be inserted	7.28
Cl 5.3	Contractor's public liability insurance: level of cover	8.9
Cl 5.4A, 5,4B and 5.4C	Insurance of the works: parties must select clause 5.4A or 5.4B, or set out their own insurance requirements under clause 5.4C	8.13
Cl 7.2	Adjudication: name the adjudicator or select an adjudicator nominating body.	10.8
Schedule 1	Arbitration: if arbitration is indicated in Article 7, select an arbitrator appointing body.	10.20

Health and safety documents

2.16 The employer and contractor are required to comply with the Construction (Design and Management) (CDM) Regulations 2015 (cl 3.9). In addition to this general obligation, clauses 3.9.1 to 3.9.4 refer to various specific obligations that arise out of the Regulations. Particularly important in regard to documentation are matters relating to the construction phase plan and the health and safety file.

2.17 The construction phase plan is not a contract document under MW16, and the recitals make no mention of it. However, the employer and the principal designer must provide the contractor with pre-construction information (regulations 4(4) and 12(3)), which should be sent out with the tender documents. Where the contractor is the principal contractor, the contractor must ensure that the construction phase plan is prepared before setting up the construction site (regulation 12); compliance with this is required under clause 3.9.2. To avoid uncertainty, it is advisable to require that this document be submitted by the contractor well in advance of start of work on site. Following commencement, the contractor must ensure that the plan is reviewed and updated on a regular basis (regulation 12(4)).

2.18 The health and safety file is principally a matter for the principal designer, who will compile it (regulation 12(5) and (6)), but there is a requirement on the contractor to provide information for this file (regulation 12(7)). Under clause 2.9 (2.10 in MWD16) the contract administrator, before certifying practical completion, must make sure that the contractor has 'complied sufficiently' with this requirement.

Execution

2.19 The articles and the contract particulars must be completed with care and the appropriate deletions made. The articles form the heart of the agreement whereby the contractor undertakes to carry out and complete the works 'in accordance with the Contract Documents' (Article 1), and in return the employer undertakes to pay the contractor 'the Contract Sum' as adjusted in accordance with the conditions (Article 2). The articles contain the attestation that must be signed by both parties. The contract may be executed under hand or as a deed (see the Note on Execution in the form, and footnote 1 to the agreement), and in the latter case the form provides for execution by the parties as individuals or as companies. (Note that the edition for use in Scotland contains special provisions for signing in accordance with the Requirements in Writing (Scotland) Act 1995.) The contract also requires that the parties sign or initial the drawings, specification, work schedules and employer's requirements (second recital, or third recital in MWD16). Although not essential it may be sensible to also initial any schedule of rates, as this is also a 'contract document'. Guidance on completing the forms can be found in the free JCT publications MW User Checklist and MWD User Checklist and in the footnotes to the contract particulars.

Use of documents

Interpretation, definitions

2.20 MW16 sets out a list of definitions in clause 1.1, some of which cross-refer to the recitals or contract particulars. Clause 1.3 contains provisions regarding interpretation, including a gender neutral clause, and that references to a person include any individual, firm, partnership, company, etc. The headings, footnotes and guidance notes are stated not to affect the interpretation of the contract. Clauses 1.4 and 1.6 also restate the requirements of the HGCRA 1996 (as amended) concerning the serving of notices and the calculation of periods of days. Clause 1.5 excludes the rights of third parties to bring actions to enforce the terms of the contract (which they might otherwise have under the Contracts (Rights of Third Parties) Act 1999). Clause 1.8 states that the law of contract will be English law. Clause 1.6.1 requires that all notices, certificates and other communications must be in writing. Clause 1.7.1 is new to the 2016 edition, and requires that consents and approvals shall not be unreasonably delayed or withheld.

Priority of contract documents

2.21 Clause 1.2 states that the agreement and conditions are to be read as a whole, and that nothing contained in the contract drawings, the contract specification or the work schedules (or, in MWD16, the employer's requirements) 'shall override or modify the Agreement or these Conditions'. If this clause were not included, the position under common law would be the reverse; in other words, anything specifically agreed and stated in a document would normally override any standard provisions.

2.22 If the parties wish to agree special terms that differ in any way from the printed conditions, then amendments must be made to the actual form. If necessary, due to lack of space, these amendments could refer to the special terms, which could be appended to the form or included in the specification. However, amending standard forms is unwise without expert advice, as the consequential effects are difficult to predict. Deleting the second sentence of clause 1.2 could be particularly unwise as it may have unintended effects on other parts of the contract.

2.23 In theory, additional provisions that supplement but do not override the printed conditions would be binding. It might be wise, though, especially where the supplementary condition is particularly significant, to make specific reference to it in the printed form.

Inconsistencies, errors or omissions

2.24 Apart from the priority given to the printed conditions, there is no statement about which of the other documents used will take priority, should a conflict arise. Clause 2.4 (2.5 in MWD16) requires inconsistencies in or between the contract documents to be corrected, including in the case of MWD16 the employer's requirements, and any such correction is treated as a variation. The normal practice is for the contract administrator to issue an instruction regarding the correction, although the contract does not specifically require this. The contractor is required to correct any inconsistencies in the documents it prepares for the CDP works 'after the Architect/Contract Administrator has approved the manner in which the Contractor proposes to deal with the inconsistency' (cl 2.5.2). A reasonable objection would be that the correction would result in the work not complying with the employer's requirements. Because of clause 2.1, the proposal should be put forward at least seven days prior to the intended date of starting the relevant work (see paragraph 5.16).

2.25 If the contractor finds any divergence between the contract documents (including in the case of MWD16 the employer's requirements), or any instruction of the contract administrator, and statutory requirements then the contract administrator must be given immediate written notice (cl 2.5.1, or 2.6.1 in MWD16). Once a divergence is discovered, the contract administrator should issue an instruction to clarify the situation and, as above, this instruction would be treated as a variation. Provided the contractor has complied with its duty to notify, the contractor would not be liable if the works do not comply with statutory requirements.

2.26 Except in the case of divergences from statutory requirements, the contractor is not under any express obligation to point out any inconsistencies that it finds within or between the contract documents. The contract also does not place any obligation on the contractor to search for discrepancies, etc. However, the general duty to use reasonable skill and care would suggest some degree of observance could be expected. If the contractor fails to point out discrepancies that it notices, or should have noticed, and work has to be rebuilt as a result, then it is suggested that the contractor may lose the right to extra payment or to an extension of time. This is only likely to be the case where the matter is so obvious that no competent contractor could have missed it, and is unlikely to apply to errors within the employer's requirements, including divergences from statutory requirements, as MWD16 makes it clear that the contractor is not responsible for their contents (cl 2.1.4).

Custody and control of documents

2.27 MW16 contains no provisions as to custody and control of documents. The original signed contract documents would normally remain in the custody of the contract administrator, who should store them safely and retain working copies for reference throughout the life of the contract and beyond. It is normal practice to provide the employer and the contractor with certified copies.

2.28 The documents provided should not be used for any purpose other than the works, and the details of the rates or prices should not be divulged. The contractor should be required to keep one copy of all of the contract documents and information issued by the contract administrator on site at all reasonable times. All drawings, etc., which bear the name of the contract administrator should be returned upon final payment.

Sub-contract documents

2.29 JCT Ltd publishes a generic short form of sub-contract (ShortSub16) that may be used with MW16, although MW16 does not require that it is used, and in practice on small projects it is unlikely that the employer will insist on the use of a particular form. MW16 does, however, include some restrictions on the terms that may be agreed for any sub-contract. These are set out in clause 3.3.2 and state the right to interest on unpaid amounts properly due to the sub-contractor from the contractor and that the contract between contractor and sub-contractor shall terminate immediately upon the termination of the contractor's employment under this contract. There are no requirements that particular conditions relating to ownership of unfixed goods and materials are included, such as those that are required by SBC16 or IC11,[1] which could in some circumstances leave the employer at risk. JCT Ltd also publishes a minor works sub-contract with sub-contractor's design (MWSub/D) for use with MWD16. Any sub-contract should also, of course, comply with the relevant requirements of the HGCRA 1996 (as amended), otherwise the terms of the Scheme for Construction Contracts would be implied.

1 At the time of writing this Guide, the 2016 edition of the JCT Intermediate Building Contract was in development. Specific references to that form are therefore to the 2011 edition.

3 Obligations of the contractor

3.1 The contractor's paramount obligation is to 'carry out and complete the Works'. This is stated in Article 1, and reinforced in clause 2.1. A court would normally hold the contractor wholly responsible for achieving this, irrespective of whether the contract administrator visits the site. Clause 3.3.1 also makes it clear that this obligation is not reduced in any way if some of the work is sub-contracted.

The works

3.2 The works that the contractor undertakes to carry out will be as briefly described in the first recital of MW16, and as shown or described in the contract documents. It is therefore important to check that the entry in the first recital clearly identifies the nature and scope of the proposed work, and that descriptions of the works given elsewhere are clear and adequate.

3.3 Note that the works will also include any changes subsequently brought about by a contract administrator's instruction, which might also introduce additional drawings or other information. These might not be 'Contract Documents', but they nevertheless have an important status and the contractor is obliged to carry out any additional work that they show.

Design

3.4 MWD16 makes provision for the contractor to undertake the design of part or parts of the project. This is brought about by the second recital (MWD16 only), which allows the parties to agree that the 'Works' will include the design and construction of identified parts, termed the 'Contractor's Designed Portion' (the 'CDP'). It will be important to identify the parts precisely (these are to be listed in the second recital or, more likely, in a separate document referenced in the recital). If this provision is to be used, the arrangement must be clearly explained to the employer, preferably at appointment stage, and confirmed in writing, as the contract administrator has no authority to delegate design responsibilities without the agreement of the employer (*Moresk Cleaners* v *Hicks*).

Moresk Cleaners Ltd v *Thomas Henwood Hicks* (1966) 4 BLR 50

Moresk Cleaners employed Thomas Hicks, an architect, to prepare plans and specifications for an extension to their laundry. Although the building was built according to the plans and specifications, the design of the structure had in fact been delegated by the architect to the contractor. Within two years cracks appeared in the structure and the roof purlins sagged. Moresk Cleaners brought a claim against the architect, who argued that it was an implied term of his contract that he should be able to delegate the design to the contractor, or alternatively that he had authority to employ the contractor on behalf of Moresk. The court found that there was no such implied term and that the architect had no such authority.

3.5 The contractor's design obligation is set out under clause 2.1.1, which states the contractor 'using reasonable skill, care and diligence, shall complete the design for the Contractor's Designed Portion, including, so far as not described or stated in the Employer's Requirements, the selection of any specifications for the kinds and standards of the materials, goods and workmanship to be used in the CDP Works'.

3.6 This level of design liability is less than that which would normally be implied by law, i.e. a strict obligation to supply something fit for purpose, but is instead an obligation to use reasonable skill and care. In effect, this means that in order to prove a breach the employer would need to prove that the contractor had been negligent. If, for example, the contractor is required to design a heating system to heat the rooms to a certain temperature, and when installed it fails to do so, this fact alone would not be enough to prove that there had been a breach of contract. The employer would need to prove that the contractor had failed to use the required level of skill and care.

3.7 Clause 2.1.1 also uses slightly different wording to the equivalent clauses in other JCT forms, which require the contractor to use the skill and care of 'an appropriately qualified and competent professional designer'. MWD16 instead requires 'reasonable skill, care and diligence'. Whereas in many contexts these may be the same level, in some a 'reasonable' level might be less. For example, on a project using a small firm of builders, it might not be reasonable to expect the same level of design skill as would be provided by a competent professional designer.

3.8 In drafting these clauses the JCT has taken a practical approach, which reflects the reality for smaller projects, where more stringent requirements may result in unacceptably high tenders. However, it is important that the contract administrator and employer understand the implications of the clause, and if a higher level of liability is required, the employer may need to consider using another form, or take legal advice.

3.9 MWD16 clause 2.1.4 makes it clear that the contractor is not responsible for the adequacy of the information in the employer's requirements. Where the employer's requirements contain an outline design, which the contractor is required to complete, the contractor is not responsible for checking the adequacy of the design that has been provided. If any inadequacy is found in any design contained in the requirements, then the contract administrator would need to issue an instruction to remedy this. The contractor is required to comply with directions of the contract administrator with regard to the integration of the design of the CDP with the works as a whole (cl 2.1.2) and, although the contract does not specifically cover the point, it is generally assumed that the contract administrator is responsible for the overall co-ordination of the design.

Materials, goods and workmanship: MW16

3.10 Under MW16 work must be carried out in a workmanlike manner and in accordance with the construction phase plan (cl 2.1.1). Under clause 2.1.1 (MW16 only) the contractor is obliged to carry out the works in compliance with the contract documents, which would include providing materials, workmanship, etc. to the standards specified. This obligation is not qualified by use of the phrase 'so far as procurable', as is the case in some other standard forms. If a specified material proved to be unavailable at the time required, it is suggested that the contractor would remain under an obligation to provide a close substitute, subject to the contract administrator's approval. If this were to constitute a variation, the contractor may then, however, be entitled to an extension of time and to an adjustment to the contract sum as a consequence. Failure to comply can be grounds for termination by the employer under clause 6.4.1.2.

3.11 Clause 2.1.2 states:

> Insofar as the quality of materials or standards of workmanship are stated to be a matter for the Architect/Contract Administrator's approval, such quality and standards shall be to his reasonable satisfaction.

This phrase does not authorise the contract administrator to alter the standard specified at will, but means that where the contract specifically states that a matter is for the approval (or discretion or opinion) of the contract administrator, the contractor only fulfils its obligations if the contract administrator is satisfied. It is suggested that clause 1.7.1 requires that any expression of dissatisfaction by the contract administrator must be made within a reasonable period of the carrying out of the work, provided it is requested by the contractor. The contract administrator should generally avoid using phrases such as 'to approval' or 'to the contract administrator's satisfaction' in the contract documents, as it leaves much scope for argument about what standard might be reasonable. It should be noted, however, that in MW16 no certificate subsequently issued by the contract administrator is stated to be conclusive as to the contract administrator's satisfaction with such work (see paragraphs 7.30–7.31).

3.12 If the phrase 'or otherwise approved' is used in a specification or bill of quantities, this does not mean that the contract administrator must be prepared to consider alternatives put forward by the contractor, nor that the contract administrator must give any reasons for rejecting alternatives (*Leedsford* v *City of Bradford*). It merely gives the contract administrator the right to do so. A substitution would always constitute a variation, whether or not this phrase is present in the specification.

***Leedsford Ltd* v *The Lord Mayor, Alderman and Citizens of the City of Bradford* (1956) 24 BLR 45 (CA)**

In a contract for the provision of a new infant school, the contract bills stated 'Artificial Stone ... The following to be obtained from the Empire Stone Company Limited, 326 Deansgate, or other approved firm'. During the course of the contract the contractor obtained quotes from other companies and sent them to the architect for approval. The architect, however, insisted that Empire Stone was used and, as Empire Stone was considerably more expensive, the contractor brought a claim for damages for breach of contract. The court dismissed the claim, stating 'The builder agrees to supply artificial stone. The stone has to be Empire Stone unless the parties agree some other stone, and no other stone can be substituted except by mutual agreement. The builder fulfils his contract if he provides Empire Stone, whether the Bradford Corporation want it or not; and the Corporation Architect can say that he will approve of no other stone except the Empire Stone' (Hodson LJ at page 58).

Materials, goods and workmanship: MWD16

3.13 Under MWD16 the clauses that deal with standard of work and materials are somewhat different from those under MW16. The contractor is required to carry out the works in compliance with the contract documents (cl 2.1), so for work which is not part of the CDP, the contractor would be obliged to comply with any specification provided. The form also contains an equivalent provision regarding work which is to be to the approval of the contract administrator (cl 2.2.1 in MWD16).

3.14 With respect to the CDP, the contractor must provide materials, goods, etc. as specified in the employer's requirements or, if none are specified, use reasonable skill and care in selecting such material and goods. Work required to be 'to approval' is similarly covered by clause 2.2.1. Clause 2.2.1 of MWD16 then continues:

> To the extent that the quality of materials and goods or standards of workmanship are neither described … nor stated to be a matter for such approval or satisfaction, they shall in the case of the Contractor's Designed Portion be of a standard appropriate to it and shall in any other case be of a standard appropriate to the Works.

This reflects the duty that would normally be implied by law; in other words, where the description of the standard required for any goods, materials and workmanship is (deliberately or inadvertently) incomplete, the contractor is required to provide something 'fit for purpose'. This appears to be a strict obligation (see above), rather than an obligation to use reasonable skill and care. There is therefore something of a tension between clauses 2.1.1 and 2.2.1 of MWD16. It is suggested that the correct interpretation is that the contractor is under a strict liability to provide materials, goods, etc. of an appropriate quality for non-CDP items, and is under the lesser obligation to use due skill and care regarding all other design forming part of the CDP, but the exact interpretation of these clauses may need to be determined by the courts. It is interesting to note that the clause quoted above does not appear in MW16, therefore that form does not cover the situation where the specification of a material, etc. is omitted. For MW16 it will be open to argument what level of responsibility, if any, the contractor has with respect to such missing items.

3.15 In summary, the contractor's responsibility for work and materials in the two forms is as follows:

- if specified or described in the contract documents, to provide materials, etc. as specified;
- if the contract requires them to be approved by the contract administrator, to be to the contract administrator's reasonable satisfaction;
- if not specified at all, with respect to the CDP, to be of a standard appropriate to the CDP (with some ambiguity over the level of liability);
- if not specified at all, with respect to non-CPD items, in the case of MWD16, to be of a standard appropriate to the works (with some ambiguity over the level of liability);
- if not specified at all, in the case of MW16, the obligations of the contractor will depend on the particular provisions agreed and whether a term may be implied regarding design liability.

Obligations in respect of quality of sub-contracted work

3.16 It is common practice today, even on small jobs, for much building work to be sub-contracted. This arrangement benefits the employer, who has the advantage of access to a wider range of skills than is normally found in the traditional building company, without any of the problems of direct responsibility for appointment of sub-contractors.

3.17 MW16 provides for domestic sub-contractors only, and there are no provisions for naming or nominating firms selected by the employer or contract administrator. The contract makes it clear that the contractor remains fully responsible for the standards and quality of all sub-contracted work, which would include any part of any CDP which had been sub-contracted (cl 3.3.1).

Compliance with statute

3.18 The contractor is under a statutory duty to comply with all legislation that is relevant to the carrying out of the work; for example, in respect of goods and services, building and construction regulations and health and safety. The duty is absolute and there is no possibility of contracting out of any of the resulting obligations.

3.19 Both MW16 and MWD16 introduce a contractual duty in addition to the statutory duty, which provides protection to the employer. Clause 2.1.1 (2.1 in MWD16) requires the contractor to comply with and give all notices required by the 'Statutory Requirements', which are defined as including any statute, statutory instrument, regulation, rule or order or any bye-law applicable to the works. As discussed in Chapter 2, if the contractor finds any divergence between the contract documents – including in the case of MWD16 the employer's requirements – or any instruction of the contract administrator and statutory requirements then it must notify the contract administrator immediately (cl 2.5.1, or 2.6.1 in MWD16), who should then issue an instruction to clarify the situation. Provided it complies with its obligation to notify the contractor administrator, the contractor is not liable for work which does not comply with statute if the non-compliance resulted from carrying out work in accordance with the contract documents or further instructions issued by the contract administrator (cl 2.5.2, or 2.6.2 in MWD16). The contractor is required to pay all fees and charges 'legally demandable' (cl 2.6, or 2.7 in MWD16), and such payments are stated not to be reimbursable, unless otherwise agreed.

Health and safety legislation

3.20 The appropriate deletion in the contract particulars (fifth recital, or sixth recital in MWD16) should indicate whether or not the project is notifiable under the CDM Regulations 2015. Where it is, the employer would be obliged to give statutory notice to the Health and Safety Executive 'as soon as practicable' before construction work begins (regulation 6(2)).

3.21 Under clause 3.9 each party undertakes to the other to comply with all their obligations under the Construction (Design and Management) (CDM) Regulations 2015. The employer must appoint a principal designer and a principal contractor (regulation 5(1); unless there is only one 'contractor' on site, which would be very unlikely in the context of MW16). The contract states that the principal designer is the contract administrator or some other person who is to be named in Article 4. Similarly, the principal contractor is the contractor unless otherwise stated (Article 5). The employer will need to be sure that the contract administrator and contractor are willing to take on these roles, and that they have the necessary skills – if not, other arrangements will need to be made. If the employer later appoints another person to act as principal designer or principal contractor, it should notify the contractor accordingly (cl 3.9.4). It should be noted that in domestic projects, some of the client's duties are transferred to the contractor under the Regulations (or principal designer, if appointed to carry these out; regulation 7(1)). If the employer fails to appoint a principal designer, the employer is required to take on that role (regulation 5(3)), except in the case of domestic projects, where it will fall to the designer (regulation 7(2)a), whether or not it is one of the designer's duties under its appointment.

3.22 Clause 3.9.1 places a contractual obligation on the employer to ensure that the principal designer carries out his or her duties under the Regulations. This is a wider obligation than the 'must take reasonable steps to ensure that' obligation imposed by the Regulations (regulation 4(6)a). Breach of it gives the contractor the right to terminate the contract under clause 6.8.1.3. What is more likely to happen is that the contractor will claim for an extension of time. An example might be where a principal designer delays in commenting on a contractor's proposed amendment to the construction phase plan and progress is thereby delayed.

3.23 The contract assumes that the contractor will act as principal contractor, unless another firm is named (Article 5). Clause 3.9 places a duty on the contractor, if acting as the principal contractor, to comply with all the relevant duties set out in the CDM Regulations. These would include, for example, that the contractor must draw up and review the construction phase plan (regulation 12) and generally co-ordinate matters relating to health and safety during the construction phase (regulation 13).

3.24 Breach of this duty is grounds for termination under clause 6.4.1.3. A specified default notice (cl 6.4.2) still has to be given, and JCT Practice Note 27 suggested that this provision should only be used in situations where the Health and Safety Executive is likely to close the site.

3.25 The contractor should take the cost of developing the construction phase plan into account at tender stage, and no claims should be entertained for adjusting it to suit the contractor's or sub-contractor's working methods. If alterations are needed as a result of an instruction requiring a variation, then the costs would be included in the valuation of the variation and the alterations should be taken into account in assessing any application for extension of time.

Other obligations

3.26 In addition to the major obligations outlined above, the contractor also has other obligations arising out of the contract. Most significant of these are in relation to progress and programming, discussed in Chapter 4, and in regard to insurance matters, discussed in Chapter 8. The contractor's obligations are summarised in Table 3.1, with the contractor's

Table 3.1 Key obligations of the contractor

Clause (MW/MWD)	Obligation of the contractor
2.1.1/2.1	Carry out and complete the works
/2.1.1	Complete the design of the CDP
/2.1.2	Comply with regulations 8–10 of the CDM Regulations and with the contract administrator's directions regarding integration of the CDP
/2.1.3	Provide drawings and information to explain the CDP
2.1.3/2.2.2	Take all reasonable steps to encourage employees, etc. to be registered under the Construction Skills Certification Scheme

Table 3.1 Key obligations of the contractor– Continued

Clause (MW/MWD)	Obligation of the contractor
2.2/2.3	Complete the works by the date for completion
/2.5.2	Correct inconsistencies between CDP documents
2.5.1/2.6.1	Notify the contract administrator of any divergence found between statutory requirements and the contract documents and any instruction of the contract administrator
2.6/2.7	Pay statutory fees and charges
2.7/2.8	Notify the contract administrator when it appears the works will not be completed by the date for completion for reasons beyond the contractor's control
2.8.1/2.9.1	Pay or allow to the employer liquidated damages for failure to complete by the date for completion
2.10/2.11	Make good defects which appear within the rectification period
3.2	Keep a person in charge on the site at all reasonable times
3.3.2	Ensure that any sub-contract includes specified conditions
3.4	Forthwith carry out all instructions issued by the contract administrator
3.6.2	Endeavour to agree the value of variations with the contract administrator
3.9	Comply with the CDM Regulations
4.5.4	Give the employer a pay less notice, if intending to pay less than the sum certified in the final certificate
4.6.1	Pay simple interest to the payee on any amount not properly paid
4.8.1	Provide the contract administrator with all documentation reasonably required for calculating the final account
5.1	Indemnify the employer for any expense, liability, loss, claim or proceedings in respect of injury to or death of any person
5.2	Indemnify the employer for any damage to property
5.3	Take out effective insurance against its liability
5.4A	Take out insurance against loss or damage to the works as required
5.4C	Take out insurance as required in the contract particulars
5.5	Produce evidence of insurance
6.11.2	Prepare an account following termination under clauses 6.8 to 6.10
Supplemental Provision 1	Work collaboratively with other team members
Supplemental Provision 2	Establish a working environment where health and safety is of paramount concert; comply with all HSE codes; ensure personnel receive site-specific training and have access to health and safety advice; ensure there is a proper consultation with all personnel
Supplemental Provision 3	Provide details of a proposed cost saving or value improvement; negotiate to agree its value
Supplemental Provision 4	Provide the employer with information on the environmental impact of materials selected by the contractor

Table 3.1 Key obligations of the contractor– Continued

Clause (MW/MWD)	Obligation of the contractor
Supplemental Provision 5	Provide the employer with information necessary to monitor the contractor's performance against indicators; submit proposals for improvement
Supplemental Provision 6	Notify the employer of matters that may give rise to a dispute, meet and engage in good faith negotiations to resolve disputes
Supplemental Provision 7	Consent to the employer publishing any amendments to the standard form
Supplemental Provision 8	Include specific terms in sub-contracts

Table 3.2 Key powers of the contractor

Clause (MW/MWD)	Contractor's express power
3.3.1	Sub-contract the works, with the contract administrator's permission
4.4.1	Make an application for payment
4.4.2.2	Issue a payment notice
4.7.1	Suspend performance of contractual obligations
5.7	Terminate the contractor's employment
6.8	Terminate the contractor's employment
6.9	Terminate the contractor's employment
6.10	Terminate the contractor's employment
Supplemental Provision 3	Propose changes to designs and specifications
Supplemental Provision 4	Suggest amendments that may improve environmental performance

key powers set out in Table 3.2. Additional obligations also arise from the optional supplemental provisions, including the obligations to work collaboratively with all team members and to adopt best practice health and safety measures.

4 Commencement and completion

4.1 There are two paramount dates in any building contract: (1) the date for possession or commencement and (2) the date for completion of the work. It is preferable that actual dates are given at the time of tendering. Vague indications such as 'to be agreed' or 'eight weeks after approval by' should be avoided, as the start date and duration can considerably affect the tender figure, particularly in the case of smaller jobs on a tight programme. In the event that work is started without proper agreement over dates, the contract will be subject to the Supply of Goods and Services Act 1982 (section 14) or the Consumer Rights Act 2015 (section 52), which state that completion is to be within a reasonable time.

4.2 The MW16 contract particulars require a 'Works commencement date' and a 'Date for Completion' (cl 2.2, or 2.3 in MWD16). However, as printed, MW16 does not allow for either commencement or completion in stages. If work needs to be carried out in contractually binding phases, then it would be necessary to set this out in detail in the contract documents and to make amendments to the form in several places. In practice it may be simpler to use a form such as IC16, which allows for sectional completion.

Commencement by the contractor

4.3 MW16 does not refer to the contractor being given 'possession' of the site, but states simply that 'The Works may be commenced' on the date stated in the contract (cl 2.2, or 2.3 in MWD16). Such wording is a realistic reflection of what much small work entails, particularly where work may have to be adjusted because the employer expects to be in part occupancy during building operations. It would be implied, however, that the contractor should be given such access as is necessary to complete the works within the contract period, and that failure to do so would constitute a reason beyond its control, thus entitling the contractor to an extension of time. In some circumstances this may include access not just to the building where the work is to be carried out, but also to other areas in the control of the client (see *The Queen in Rights of Canada* v *Walter Cabbott Construction Ltd*). It should also be noted that the use of the word 'may' means that, in contrast to other contracts, such as the RIBA forms, the contractor is not in breach if it fails to commence work on this date. Any prolonged delay to the start could, however, constitute grounds for termination (see paragraph 9.7).

The Queen in Rights of Canada v *Walter Cabbott Construction Ltd* (1975) 21 BLR 42

This Canadian case (Federal Court of Appeal) concerned work to construct a hatchery on a site (contract 1), where several other projects relating to ponds were also planned (contracts 3 and 4). The work to the ponds could not be undertaken without occupying part of the hatchery site.

> Work to the ponds was started in advance of contract 1, causing access problems to the contractor when contract 1 began. The court confirmed (at page 52) the trial judge's view that 'the "site for the work" must, in the case of a completely new structure comprise not only the ground actually to be occupied by the completed structure but so much area around it as is within the control of the owner and is reasonably necessary for carrying out the work efficiently'.

4.4 If the work is substantially suspended for a period of a month or more, due to failure to allow necessary access, then this would be grounds for termination under clause 6.8.2.2. Failure to give the contractor access, or granting inadequate access only, may also amount to repudiation, giving the contractor the right to treat the contract as at an end or bring a claim for damages (*Whittal Builders* v *Chester-le-Street DC*). Any restrictions on access or working methods should therefore be clearly stated in the contract documents.

> ***Whittal Builders Co. Ltd* v *Chester-le-Street District Council* (1987) 40 BLR 82**
>
> Whittal Builders contracted with the Council on JCT63 to carry out modernisation work to 90 dwellings. The contract documents did not mention the possibility of phasing, but the Council gave the contractor possession of the houses in a piecemeal manner. Even though work of this nature was frequently phased, the judge nevertheless found that the employer was in breach of contract for not giving the contractor possession of all 90 dwellings at the start of the contract, and the contractor was entitled to damages.

4.5 The parties are, of course, always free to renegotiate the terms of any contract. Therefore, if there is a delay in allowing access, the parties may have to agree new dates for commencement and completion, usually with financial compensation to the contractor.

4.6 It should be noted that clause 2.2 (2.3 of MWD16) simply states that the works may be commenced on the stated date. Therefore, the contractor would not be in breach of contract if the works were commenced some time after this date. In practice this might cause the employer problems with respect to security or insurance. If the contractor is required to be responsible for the site from the stated date then suitable provisions would have to be included in the contract documents and an amendment made to the clause.

Occupation by the employer

4.7 In the context of MW16 many employers may wish to use or occupy part of the works during the time that the contractor is on site. If this is the case it should also be made clear in the tender documents, and agreement should be reached with the contractor over suitable arrangements. This situation may occur, for example, in domestic projects, or during refurbishment work to existing buildings. If at all possible, the work should be phased so that the contractor has exclusive access to certain parts for pre-agreed periods. The contract administrator may need to take a proactive role in discussing with the employer how this phasing will work prior to the tender documents being sent out. In some cases it will be inevitable that the employer will need access to the site throughout the project; for example, where repair work is carried out to the entrance hall or circulation routes in a residential block and it is not feasible to evacuate the block. Again, the requirements must be made clear in the tender documents, so that the contractor can allow for adequate health and safety measures to be taken and include any screening or temporary work, etc. in its tender. In all such cases the insurers of the works or existing buildings should be kept informed.

Progress

4.8 Although MW16 does not contain an express requirement that the contractor should work with due diligence, depending on the circumstances it is normally implied in a construction contract that a contractor will proceed 'regularly and diligently'. The contractor is free to organise its own working methods and sequences of operations, with the important qualification that it must comply with the construction phase plan. Short periods of inactivity would not necessarily constitute a breach of contract; however, failure to proceed regularly and diligently is a ground for terminating the contract (cl 6.4.1.2, see paragraph 9.7).

4.9 MW16 does not require the contractor to produce a programme, although there would be nothing to prevent such a requirement being included in the specification; a programme might be very useful to the contract administrator, particularly when monitoring progress and assessing extensions of time. It might be helpful if the contractor sets out dates by which key information will be needed, although with many small jobs this should not be necessary. The contract administrator should be careful never to 'approve' any programme in such a way that it becomes a contractually binding document against which the contract administrator's own performance in providing information will be judged.

4.10 Where a contractor's programme shows a large float period between the contractor's estimated completion date and the date for completion, this should always be queried. For example, if the contractor has tendered to carry out the work in eight months but produces a programme showing estimated completion in six months then this may be of little assistance for assessing progress and extensions of time. Also, an early completion date could create difficulties for the employer. What would not be acceptable is for the contractor to claim loss and expense on the grounds that it was prevented from achieving an early completion date through delayed information or decisions by the contract administrator. Just because a contractor's programme shows an intention to complete early, there is no implied duty on the employer to enable the contractor to achieve this early completion (*Glenlion Construction* v *The Guinness Trust*).

Glenlion Construction Ltd v *The Guinness Trust* (1987) 39 BLR 89

The Trust employed Glenlion Construction on JCT63 to carry out works in relation to a residential development at Bromley, Kent. The contract required the contractor to provide a programme. Disputes arose which went to arbitration, and several questions of law regarding the contractor's programme were subsequently raised in court. The contractor claimed loss and expense on the ground that it was prevented from economic working and achieving the early completion date shown on its programme because the architect had failed to provide necessary information and instructions at the dates shown. The court decided that Glenlion was entitled to complete before the date for completion and entitled to carry out the works in a way which would achieve an earlier completion date. However, there was no implied obligation that the employer (or the architect) should perform its obligations so as to enable the contractor to complete by any earlier completion date shown on the programme.

Completion

4.11 Building contracts usually refer to only one completion date for the works. The significance of having a completion date is that it provides a fixed point from which damages may be payable in the event of non-completion. Generally in construction contracts the damages are 'liquidated', and usually expressed as a rate per week of overrun.

4.12 The contractor is obliged to complete the works by the completion date, and in general accepts the risk of any events that might prevent completion by this date. The contractor is relieved of this obligation if the employer causes delays or in some way prevents completion. Most contracts contain provisions allowing for the adjustment of the completion date in the event of certain delays caused by the employer, or in the event of neutral delaying events. The contract dates can, of course, always be adjusted by agreement between the parties.

4.13 It is sometimes essential that completion is achieved by a particular date and failure would mean that the result is worthless. This is usually expressed as 'time is of the essence'. Breach of such a term would be considered a fundamental breach, and the employer would be entitled to terminate performance of the contract and treat all its own obligations as at an end. However, an expression such as 'time is of the essence' is seldom applicable to building contracts, because the inclusion of extensions of time and liquidated damages provisions imply that the parties intended otherwise (*Gibbs* v *Tomlinson*). Nevertheless, there could be occasions where MW16 is used for short-term exhibition or special event constructions, and where failure to meet the date of the event will result in a total loss of the intended benefit to the employer, in which case 'time is of the essence' might be a realistic requirement. In such circumstances the wording of the contract would need to be amended – always, of course, with the benefit of legal advice.

Gibbs v *Tomlinson* (1992) 35 Con LR 86

Mr Gibbs employed Tomlinson on the Agreement for Minor Building Works (MW80) to carry out alterations and construct an extension to his house. Work commenced on site and proceeded satisfactorily until about December 1989, when there were disagreements between the plaintiff and the defendant which resulted in the termination of the contract; the builder did no work on site after about 12 December 1989. The plaintiff then engaged various workmen to complete the works, some of whom had previously been employees or sub-contractors of the defendant. He then brought a claim for damages and return of monies allegedly overpaid, claiming that the contractor's failure to complete by the original completion date entitled him to treat the contract as at an end. Mr Recorder Harvey QC found this was not the case, stating that MW80 did not expressly make time of the essence, and that clause 2.3 envisaged that the contractor would continue with the work and pay liquidated damages at the specified rate.

4.14 MW16 requires an entry in the contract particulars (cl 2.2, or 2.3 in MWD16) which gives the date for completion. The contract gives the contract administrator the power to extend time, but makes no provision for the contract administrator to reduce the contract period, even if substantial work is omitted. If Supplemental Provision 3 is incorporated, however, a new date (either earlier or later) might be agreed as part of a cost-savings proposal. The contractor is obliged to complete the works by the date for completion, or any later date consequent upon an extension of time. If the contractor fails to complete by this date, liquidated damages become payable (see Figure 4.1). Once completion is attained, the

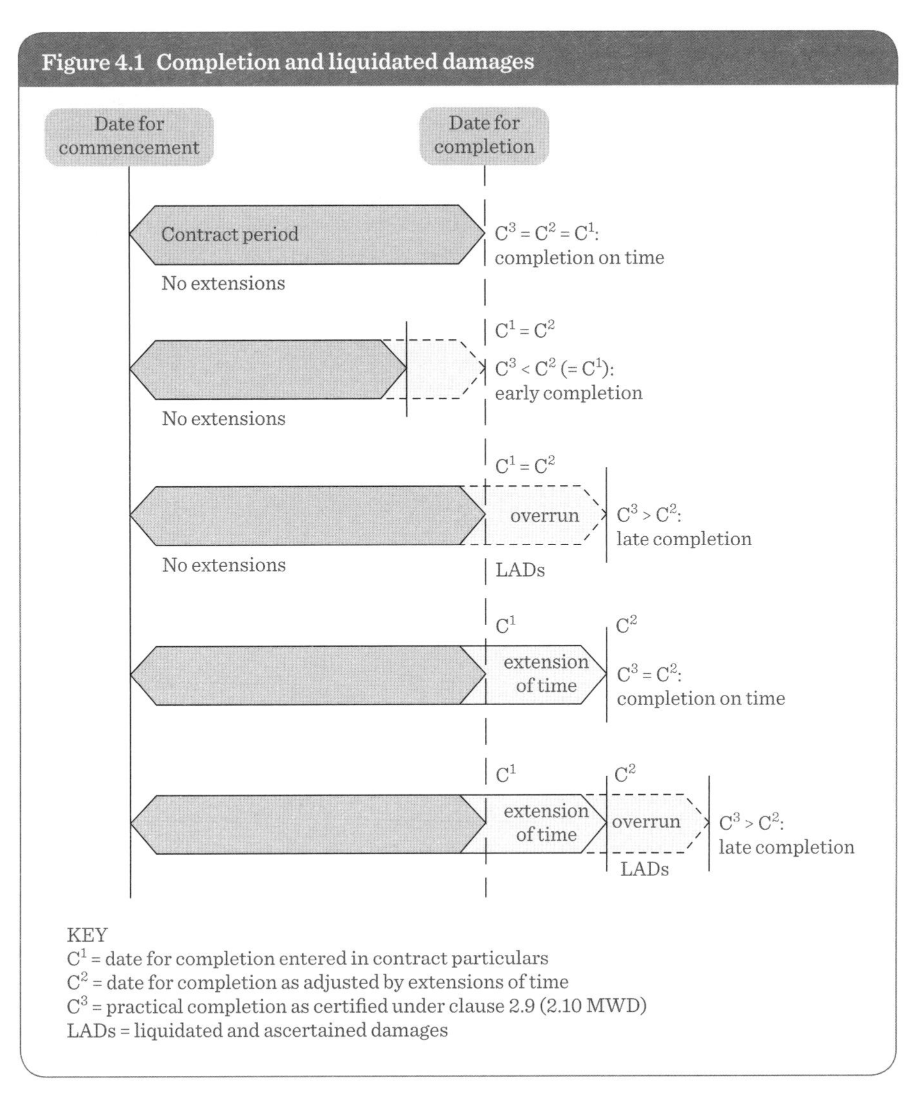

Figure 4.1 Completion and liquidated damages

practical completion certificate must be issued and the employer is obliged to accept the works. Employers who wish to accept the works only on the date in the contract would need to amend the wording.

Extensions of time

Principle

4.15 One important reason for an extension of time clause is to preserve the employer's right to liquidated damages in the event that the contractor fails to complete on time due wholly or in part to some action for which the employer is responsible. If there were no provisions to grant extensions of time and a delay occurred that was caused at least in part by the employer, this would in effect be a breach of contract by the employer and the contractor would no longer be bound to complete by the completion date (*Peak Construction* v *McKinney Foundations*). The employer would therefore lose the right to liquidated damages, even though much of the blame for the delay might rest with the contractor. The phrase 'time at large' is often used to describe this situation. In most cases, however, the contractor would nevertheless remain under obligation to complete within a reasonable time.

Peak Construction (Liverpool) Ltd v *McKinney Foundation Ltd* (1970) 1 BLR 111 (CA)

Peak Construction was the main contractor on a project to construct a multi-storey block of flats for Liverpool Corporation. The main contract was not on any of the standard forms, but was drawn up by the Corporation. McKinney Foundations Ltd was the sub-contractor nominated to design and construct the piling. After the piling was complete and the sub-contractor had left the site, serious defects were discovered in one of the piles and, following further investigation, minor defects were found in several other piles. Work was halted while the best strategy for remedial work was debated between the parties. The city surveyor did not accept the initial remedial proposals, and it was agreed that an independent engineer would prepare an alternative proposal. The Corporation refused to agree to accept his decision in advance, and delayed making the appointment. Altogether it was 58 weeks before work resumed (although the remedial work took only six weeks), and the main contractor brought a claim against the sub-contractor for damages.

The Official Referee at first instance found that the entire 58 weeks constituted delay caused by the nominated sub-contractor and awarded £40,000 damages for breach of contract, based in part on liquidated damages, which the Corporation had claimed from the contractor. McKinney appealed, and the Court of Appeal found that the 58-week delay could not possibly entirely be due to the sub-contractor's breach, but was in part caused by the tardiness of the Corporation. This being the case, and as there were no provisions in the contract for extending time for delay on the part of the Corporation, it lost its right to claim liquidated damages, and this component of the damages awarded against the sub-contractor was disallowed.

Procedure

4.16 Under MW16 the contractor must give written notice to the contract administrator 'If it becomes apparent that the Works will not be completed by the Date for Completion' (cl 2.7, or 2.8 in MWD16). The notice must be given regardless of the reason for the delay, i.e. whether it is caused by the employer, by a neutral event (such as bad weather) or by the contractor itself (this is a change since the 2005 edition of this form, where the notice was only required where delay was for reasons beyond the control of the contractor). (Figure 4.2 gives an overview of the MW16 timeline.)

4.17 This obligation is nevertheless quite limited. The phrase 'If it becomes apparent that the Works will not be completed' suggests that notice only has to be given if it is reasonably clear that completion will be delayed; i.e. a notice is not required if there is simply a current delay in progress, or even if failure to complete in time appears possible. This limitation could cause problems in practice. It is often important that the contract administrator is made aware as soon as possible of any delays in progress in order to keep the employer informed, and so that measures which might mitigate the problem can be considered. Contractors often remain optimistic that temporary delays will be overcome, but this can lead to disappointment and disputes later in the project. If it is anticipated that notification

Figure 4.2 The MW16 timeline

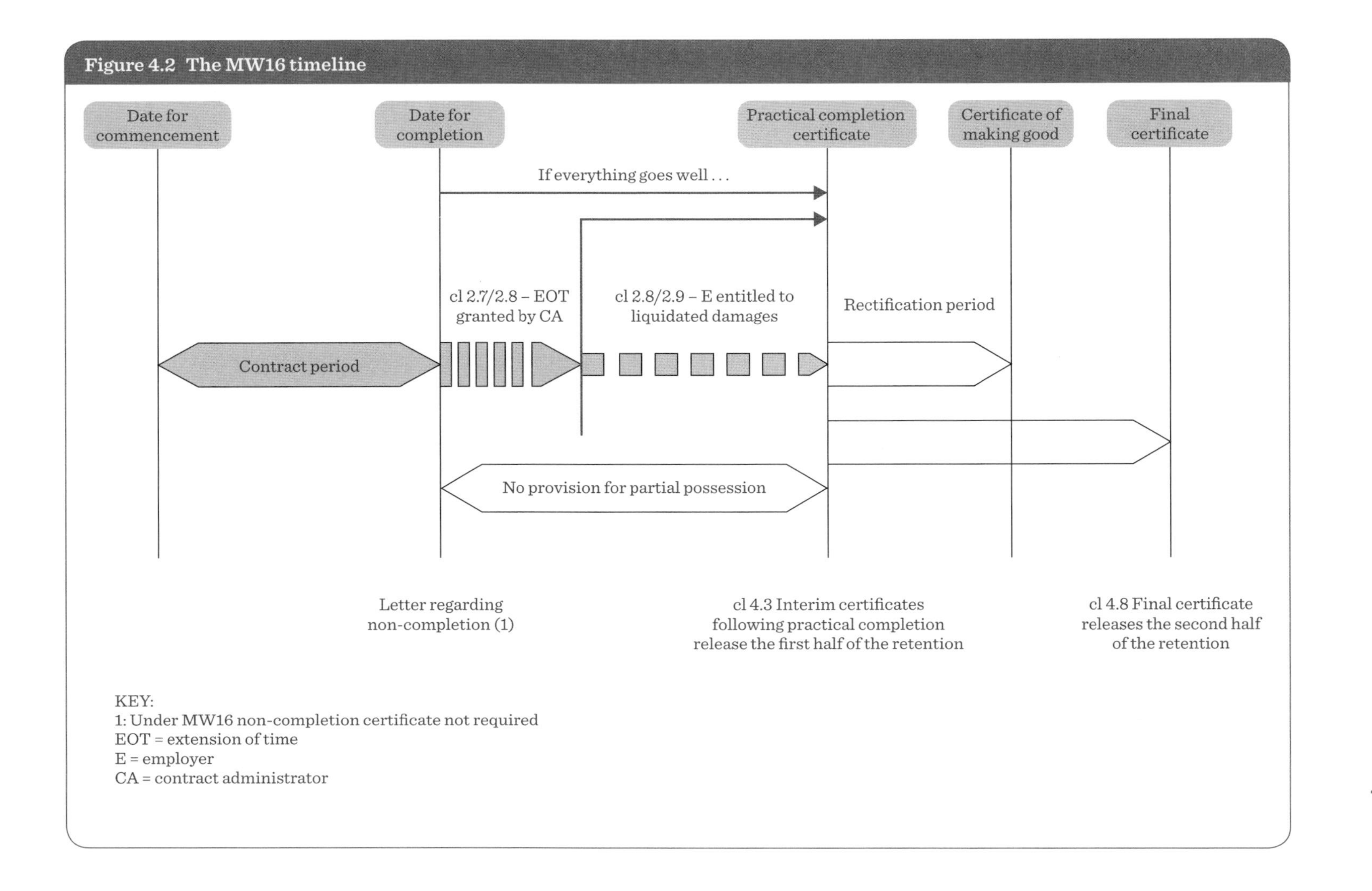

of *any* delay is essential, then this should be made clear in the tender documents and a suitable amendment made to the form.

4.18 The contract does not require the contractor to explain the reasons for the delay in the notice. However, for practical purposes it would be sensible for the contractor to give as much information as possible about the causes and the extent, so that the contract administrator can assess what extension might be appropriate.

4.19 Once notice of delay has been given 'Where that delay occurs for reasons beyond the control of the Contractor, including compliance with Architect/Contract Administrator's instructions', the contract administrator is required to make 'such extension of time for completion as may be reasonable' (cl 2.7, or 2.8 in MWD16). For guidance on what might be considered to be beyond the contractor's control, the contract administrator might have regard to those matters listed as 'Relevant Events' under SBC16, for example exceptionally adverse weather, the 'Specified Perils', strikes, failure to supply information, site access and indeed any difficulty in movement on or around site. The phrase is, however, broadly expressed and might not be limited to those events. On the other hand, the contractor could be assumed to have allowed for any circumstances which had been explained clearly in the tender documentation, such as access restrictions, even though these may in practice be beyond its control. If the nature and extent of the restrictions have been made clear then they are unlikely to entitle the contractor to an extension of time.

4.20 It is also suggested that the contractor might be expected to have allowed for any circumstances which a competent contractor could have predicted, such as the occurrence of bad (but not unusually bad) weather in January. However, this is a difficult point as it would rely on the courts implying a term that things which could be predicted are within the contractor's control. This was not the approach taken by the court in *Scott Lithgow* v *Secretary of State for Defence*, where the court preferred to give the phrase its ordinary meaning.

Scott Lithgow Ltd v *Secretary of State for Defence* (1989) 45 BLR 1 (HL)

Scott Lithgow Ltd was the successor to Scott's Shipbuilding Co. (1969) Ltd, which contracted with the Ministry of Defence to construct two 'Oberon' class submarines. The contract required that pressure-tight cables should be supplied by certain firms. One of these firms, BICC, became a sub-contractor for the supply of the cables. The cables were found to be defective and had to be replaced. Scott Lithgow brought proceedings against BICC, which were settled for a sum less than the losses suffered through having to replace the cables. Scott Lithgow then gave notice of arbitration under the main contract.

A case was then stated to the Inner House of the Court of Session, and certain points appealed to the House of Lords. Clause 20A.3 of the contract had stated that the contractor should be paid for the effect of 'exceptional dislocation and delay arising during the construction of the vessel due to alterations, suspension of work, or any other cause beyond the contractor's control'. The House of Lords held that failure by suppliers or sub-contractors in breach of their contractual duties to Scott Lithgow was not a matter which, according to the ordinary use of language, could be regarded as within Scott Lithgow's control. Lord Keith of Kinkel pointed out that whether or not the failure of the cables had been a matter within the control of Scott Lithgow was a question of fact. The contractor's position depends on demonstrating that 'he has no means of securing that (his requirements) are met. If the contractor failed to stipulate a time for delivery, consequent delay would be his own responsibility but if he did so stipulate and delivery was late the position would be different' (at page 13).

4.21 MW16 makes it clear that default of sub-contractors or suppliers (its own or its sub-contractors') is a reason within the control of the contractor (cl 2.7, or 2.8 in MWD16). It is suggested that the wording of this is clear enough to include even default by sub-contractors or suppliers named by the contract administrator in the specification.

4.22 There are no time limits on when the contract administrator must respond to the contractor's notice, but it is suggested that this should be done as soon as possible, in order to preserve the employer's right to liquidated damages. The contract administrator should either fix a new completion date or notify the contractor that no extension of time is due. The contract administrator might call for information if this is necessary to make a fair and reasonable assessment, but this must never be considered as a delaying tactic. The contract does not appear to prevent the contract administrator later awarding an additional extension, if this seems reasonable in the light of further information.

4.23 The right of the contract administrator to award an extension of time is limited in several important respects. First, it appears to be dependent on the contractor having given notice, and it is even arguable that the notice should be given before the date for completion. This will present the contract administrator with a dilemma in the unlikely event that the contractor fails to submit a notice where the employer has caused delay, or submits the notice late. The right to deduct liquidated damages may be jeopardised if the contract administrator is unable to extend the contract period. The best policy may be to award an extension in any case, allowing the contractor to take the matter to adjudication if it chooses. It would seem unlikely that a court would find that the parties had intended that absence or lateness of notice should have such a drastic effect (*Terry Pincott* v *Fur & Textile Care*).

***Terry Pincott* v *Fur & Textile Care Ltd* (1986) 3 CLD-05-14**

Fur & Textile Care, a dry cleaning business, appointed architect David Daw in relation to an extension to its premises. Terry Pincott was engaged as contractor on the Minor Works Agreement. A certificate of practical completion was issued on 13 August 1982. The contractor then submitted a request for an extension of time of 16 weeks on 8 September, which the architect granted in full on 14 September. HH Judge Smout QC stated that 'The request did not accord with the terms of clause 6(ii) of the Minor Works Agreement in that [the contractor] did not notify [the architect] prior to the anticipated date for completion. I doubt, however, whether time is of the essence in regard to that sub-clause'. The lateness of itself would probably not have invalidated the extension, but the architect in this case failed to exercise independent professional judgment in awarding it. The extension was therefore not a bona fide exercise of his powers and was invalid.

4.24 The second limitation is that there appear to be no provisions whereby the contract administrator may reduce a previous extension of time by fixing an earlier completion date where work has been omitted. Nevertheless, it is suggested that if work has been omitted, the contract administrator could take this into account when deciding what might be a reasonable extension in response to some further notice by the contractor.

4.25 The third limitation is that there appear to be no provisions whereby the contract administrator may award further extensions of time in respect of delaying events which occur after the date for completion or any extended completion date, i.e. when the contractor is in 'culpable delay'. The use of the phrase 'will not be completed by' suggests that clause 2.7 (cl 2.8 in MWD16) is referring only to events before the date for completion. The court in *Balfour Beatty* v *Chestermont Properties* concluded that the contract administrator had no such power, but this was in the context of a form that made express provision for a 'review' of extensions of time. There is no such review provision in MW16, so it may be that it would be viewed differently, but cases in the past indicate that courts take a strict approach to implying terms which extend the employer's rights in regard to extending time. The best approach may be for the employer to agree a revised completion date with the contractor, who in practice is unlikely to object as there would be little to be gained. Nevertheless, this emphasises the importance of dealing with all extension of time applications promptly.

Balfour Beatty Building Ltd v *Chestermont Properties Ltd* (1993) 62 BLR 1

In a contract on JCT80, the works were not completed by the revised completion date and the architect issued a non-completion certificate. The architect then issued a series of variation instructions and a further extension of time, which had the effect of fixing a completion date two-and-a-half months before the first of the variation instructions. He then issued a further non-completion certificate and the employer proceeded to deduct liquidated damages. The contractor took the matter to arbitration and then appealed certain decisions on preliminary questions given by the arbitrator. The court held that the architect's power to grant an extension of time pursuant to clause 25.3.1.1 (similar to clause 2.7 in MW16) could only operate in respect of relevant events that occurred before the original or the previously fixed completion date, but the power to grant an extension under clause 25.3.3 applied to any relevant event.

Assessment

4.26 It is an obligation on the contract administrator to issue extensions of time when properly due and any failure on the part of the contract administrator to do so is a breach on the part of the employer. In every case the contract administrator should assess the effect of the delay on the contract completion date. A contractor's programme could be useful as a guide, but it would not be binding. The effect on progress is assessed in relation to the work being carried out at the time of the delaying event, rather than the work that was programmed to be carried out. The contract does not set this out, but it is suggested that it would be implied that the contractor should take reasonable steps to prevent delay, although not to the extent of incurring excessive expenditure. The contract administrator is not required to give reasons as to why a particular extension was awarded. However, it would be prudent to make careful records in case the matter is taken to adjudication.

4.27 The effects of any delay on completion are not always easy to predict. Nevertheless, the contract administrator is required to reach an opinion, and in doing this owes a duty to both parties to be fair and reasonable (*Sutcliffe* v *Thackrah*, see paragraph 7.1). This applies even where the delay may have been caused by the contract administrator, for example where the contract administrator has failed to issue information in sufficient time.

4.28 It sometimes happens that two or more delaying events occur simultaneously, or with some overlap, and this can raise difficult questions with respect to the awarding of extensions of time. In the case of concurrent delays involving two or more causes, it has been customary to grant the extension in respect of the dominant reason, but this is only appropriate where the dominant reason begins before and ends after any other reasons.

4.29 Where one overlapping delaying event is beyond the control of the contractor and the other is not – in other words, one is the employer's risk and the other the contractor's –

a difficult question arises as to what extension of time is due. The instinctive reaction of many assessors might be to 'split the difference', given that both parties have contributed to the delay. However, the more logical approach is that the contractor should be given an extension of time for the full length of delay caused by the employer, irrespective of the fact that, during the overlap, the contractor was also causing delay. Taking any other approach – e.g. splitting the overlap period and awarding only half of the extension to the contractor – could result in the contractor being subject to liquidated damages for a delay partly caused by the employer. The courts have normally adopted this analysis (see *Walter Lilly & Co. Ltd* v *Giles Mackay & DMW Ltd*). A leading Scottish case had stated that a proportional approach would be fairer, but this decision is not binding on English courts and was not approved in *Walter Lilly*.

Walter Lilly & Co. Ltd v *Giles Mackay & DMW Ltd* [2012] EWHC 649 (TCC)

This case concerned a contract to build Mr and Mrs Mackay's, and two other families', luxury new homes in South Kensington, London. The contract was entered into in 2004 on the JCT Standard Form of Building Contract 1998 Edition with a Contractor's Designed Portion Supplement. The total contract sum was £15.3 million, the date for completion was 23 January 2006, and liquidated damages were set at £6,400 per day. Practical completion was certified on 7 July 2008. The contractor (Walter Lilly) issued 234 notices of delay and requests for extensions of time, of which fewer than a quarter were answered. The contractor brought a claim for, among other things, an additional extension of time. The court awarded a full extension up to the date of practical completion. It took the opportunity to review approaches to dealing with concurrent delay, including that in the case of *Henry Boot Construction (UK) Ltd* v *Malmaison Hotel (Manchester) Ltd* (where the contractor is entitled to a full extension of time for delay caused by two or more events, provided one is an event which entitles it to an extension under the contract), and the alternative approach in the Scottish case of *City Inn Ltd* v *Shepherd Construction Ltd* (where the delay is apportioned between the events). The court decided that the former was the correct approach in this case. As part of its reasoning the court noted that there was nothing in the relevant clauses to suggest that the extension of time should be reduced if the contractor was partly to blame for the delay.

Occupation before practical completion

4.30 MW16 makes no provision for the employer to use or occupy the site or the works or any part prior to practical completion. If arrangements for phased occupation have not been agreed in the contract documents, a situation may arise where the contractor has not completed by the date for completion but part of the works are complete or sufficiently complete to allow the employer to have beneficial use of those parts and the employer is anxious to occupy them. There is nothing in the contract that allows for this; therefore, a separate ad hoc agreement would have to be made.

4.31 An interesting suggestion was put forward in the 'Practice' section of the *RIBA Journal* (February 1992) (see Figure 4.3). Although not directed at MW16, it could be adapted to suit this contract. In this arrangement, in return for being allowed to occupy the premises, the employer agrees not to claim liquidated damages during the period of occupation, or to claim it at a reduced rate. Practical completion obviously cannot be certified, and there is no release of retention money until it is. Matters of insuring the works will need to be settled with the insurers. Because such an arrangement would be outside the terms of the contract it should be covered by a properly drafted agreement which is signed by both parties. It may also be sensible to agree that, in the event that the contractor still fails to

Figure 4.3 'Practice' section, *RIBA Journal* (February 1992)

Employer's possession before practical completion under JCT contracts

It is not uncommon for the employer, after the completion date has passed, to wish to take possession of the Works before the contractor has achieved practical completion. In this event an ad hoc agreement between employer and contractor is required to deal with the situation. In respect of such an agreement, members may wish to have regard to the following note . . .

Outstanding items

Where it is known to the architect that there are outstanding items, practical completion should not be certified without specially agreed arrangements between the employer and the contractor. For example, in the case of a contract where the contract completion date has passed it could be so agreed that the incomplete building will be taken over for occupation, subject to postponing the release of retention and the beginning of the defects liability period until the outstanding items referred to in a list to be prepared by the architect are completed, but relieving the contractor from liability for liquidated damages for delay as from the date of occupation, and making any necessary changes in the insurance arrangements. In such circumstances either the Certificate of Practical Completion form should not be used or it should be altered to state or refer to the specially agreed arrangements. In making such arrangements the architect should have the authority of the client-employer.

When the employer is pressing for premature practical completion there is a need to be particularly careful where there are others who are entitled to rely on the issue of a Practical Completion Certificate and its consequences. In the case where part only of the Works is ready for hand-over the partial possession provisions can be operated to enable the employer with the consent of the contractor to take possession of the completed part.

achieve practical completion by the end of an agreed period, the rate of liquidated damages would then increase. In most circumstances this arrangement would be of benefit to both parties, and is far preferable to issuing a heavily qualified certificate of practical completion listing 'except for' items.

Practical completion

4.32 Under clause 2.9 (2.10 in MWD16) the contract administrator is obliged to certify the date at which, in the contract administrator's opinion, works have reached practical completion and the contractor has complied sufficiently with clause 3.9 (its CDM obligations, for example the supply of information required for the health and safety file) and, in the case of MWD16, clause 2.1.3 (supply of CDP drawings). The date certified is the date when the last condition is fulfilled: in other words, if there is a delay before receiving the health and safety information, the date of its receipt should be the date on the certificate, irrespective of the fact that practical completion of the works was achieved days or even weeks earlier. The use of the term 'complied sufficiently' may allow the contract administrator to use its discretion in issuing the certificate with information missing. The contract administrator should, however, be very careful not to place the employer in a position where it would be in breach of the CDM Regulations.

4.33 Unfortunately, the drafting of the contract is not as consistent as it might be in relation to practical completion. It refers in several clauses to 'the date of practical completion' rather than the date certified under clause 2.9 (2.10 in MWD16). As clause 2.9 (2.10) refers to practical completion of the works *and* the provision of CDM (and CDP) information as two separate events, both of which have to have occurred for the certificate to be issued, there is at least room for argument that where other clauses refer to 'the date of practical completion' they are referring to the first event only.

4.34 For example, in clause 2.8 (2.9 in MWD16) liability for liquidated damages is stated to run until the 'date of practical completion', and a contractor may try to argue that it should cease once the works have reached practical completion. It is suggested that the correct interpretation is that the liability runs until the date certified under clause 2.9 (2.10 in MWD16), regardless of when the works reach practical completion, not least because the employer may well be in breach of its statutory obligations until the necessary information is provided. The ambiguity in the drafting arises from the failure of the JCT to adjust related clauses after the reference to clause 3.9 (CDM obligations) was added to clause 2.9 (2.10 in MWD16).

Practical completion of the works

4.35 Deciding when the works have reached practical completion often causes the contract administrator some problems. As with other decisions under the contract, it is implied that it will be a fair and reasonable exercise of professional judgment. The contract administrator should be satisfied that there are no patent defects, that all construction work as defined in the contract has been completed, and that if the CDM Regulations apply in full, the contractor has sufficiently complied with obligations in respect of the health and safety file. However, it has been held that the contract administrator has a discretion to certify practical completion where there are very minor items of work left incomplete, on *de minimis* principles (*H W Nevill (Sunblest)* v *William Press*, *Laing O'Rourke* v *Healthcare*

H W Nevill (Sunblest) Ltd v *William Press & Son Ltd* (1981) 20 BLR 78

William Press entered into a contract with Sunblest to carry out foundations, groundworks and drainage for a new bakery on a JCT63 contract. A practical completion certificate was issued, and new contractors commenced a separate contract to construct the bakery. A certificate of making good defects and a final certificate were then issued for the first contract, following which it was discovered that the drains and the hard standing were defective. William Press returned to the site and remedied the defects, but the second contract was delayed by four weeks and Sunblest suffered damages as a result. It commenced proceedings, claiming that William Press was in breach of contract and in its defence William Press argued that the plaintiff was precluded from bringing the claim by the conclusive effect of the final certificate.

Judge Newey decided that the final certificate did not act as a bar to claims for consequential loss. In reaching this decision he considered the meaning and effect of the certificate of practical completion and stated (at page 87):

> I think that the word 'practically' in clause 15(1) gave the architect a discretion to certify that William Press had fulfilled its obligation under clause 21(1) where very minor *de-minimis* work had not been carried out, but that if there were any patent defects in what William Press had done then the architect could not have issued a Certificate of Practical Completion.

***Laing O'Rourke Construction Ltd (formerly Laing O'Rourke Northern Ltd)* v *Healthcare Support (Newcastle) Ltd* [2014] EWHC 2595 (TCC)**

This case concerned a hospital project, where the relevant contract provided for the engagement of an independent tester who would certify practical completion of the claimant's works in accordance with the completion criteria specified in the project agreement. The independent tester refused to certify practical completion due to the employer's complaints about the quality and conformity of certain aspects of the works (toilet area size, daylight levels, window restrictors, link bridge steelwork and room temperature). The contractor argued that these matters fell outside the completion criteria, and that compliance with the criteria was sufficient. The court agreed with this view. However, in reaching its opinion it acknowledged that, where there are departures from the criteria, the tester would nevertheless be justified in issuing the certificate where the departure would not have any material adverse impact on the ability of the employer to enjoy and use the buildings for the purposes anticipated by the contract. There was no justification to imply a term that any breach of the contractual requirements, however technical or minor, would prevent certification of practical completion.

Support). It has also been held that it is sufficient that the contractor complies with the technical standards and requirements set out in the contract; it does not have to satisfy any other unstated general employer requirements (*Laing O'Rourke* v *Healthcare Support*).

4.36 However, such discretion should be exercised with extreme caution. It should not extend to the issue of a certificate qualified by 'except for the following items' followed by a long list. The rectification period will commence, the contractor may feel little incentive to return, particularly if the retention money held hardly justifies the cost, and the client is likely to take an exaggerated view of what defects remain. It may be worth reminding any employer that is pressing for a certificate of practical completion of the consequences of such action.

4.37 A key consequence is that half of the retention is released, which leaves a retention of only 2.5 per cent (or half of the percentage stated in the contract particulars) in hand. The money may be used to remedy work which the contractor refuses to correct, but is only intended to cover the risk of latent defects, and may not be enough to cover defects which are apparent at practical completion.

4.38 There are a number of other consequences of practical completion. First, the rectification period begins (cl 2.10, or 2.11 in MWD16). Any work which is completed during the rectification period will not have the benefit of the full period to allow latent defects to emerge. This may be particularly important with respect to services, which require a seasonal cycle to be properly tested. Second, the employer takes over responsibility for the site, and the contractor will no longer cover the insurance of the works under clause 5.4A (and possibly, depending on the arrangements made, under 5.4C). The insurers will therefore need to be informed about the programme for the outstanding works. Third, the contractor's liability for liquidated damages ends (cl 2.8.1, or 2.9.1 in MWD), so if the employer suffers further losses due to the contractor having to return to site it may find these very difficult to recover. Finally, the employer will be the 'occupier' for the purposes of the Occupiers' Liability Act 1957, and also may be subject to claims regarding health and safety.

Procedure at practical completion

4.39 The contract sets out no procedure for dealing with practical completion, it simply requires the contract administrator to certify it. The contract documents may set out a procedure, but the contract administrator should check carefully at tender stage to ensure that the procedure is satisfactory.

4.40 It appears to be widespread practice for contract administrators to issue 'snagging' lists in the period leading up to practical completion, sometimes in great detail and on a room-by-room basis. The contract does not require this, and neither do most standard terms of appointment. Under the contract, responsibility for quality control and snagging rests entirely with the contractor. In adopting the practice of 'snagging' the contract administrator might be helping the contractor and, although this may appear to benefit the employer, it can create confusion over the liability position, which could cause problems at a future date.

4.41 It is frequently the practice for the contractor to arrange a 'handover' meeting. The term is not used in MW16, and although handover meetings can be of use, particularly in introducing the finished project to the employer, it is better to avoid complex or inflexible procedures in the contract documents. Where a handover meeting has been arranged, or the contractor has stated in writing that the works are complete, it still remains the contract administrator's responsibility to decide whether practical completion has been achieved. If the contract administrator feels that the works are not complete, there is no obligation to justify this opinion with schedules of outstanding items. The best course may just be to draw attention to typical items, but to make it clear that any list is indicative and not comprehensive.

4.42 The 2016 edition of the contract has removed the special provisions covering a 'penultimate' certificate. Instead, interim certificates will continue to be issued at the same intervals after practical completion as they had before, and following the same procedure (see paragraphs 7.3–7.6). The only difference is that, as noted above, the amount of withheld retention is cut to 2.5 per cent.

Failure to complete by the completion date

4.43 In the event of failure to complete, the employer, provided that it has issued the necessary notices, may deduct damages from the amount due under the next certificate or reclaim the sum as a debt (cl 2.8.2, or 2.9.2 in MWD16). Note that fluctuations in relation to contribution, levy and tax changes are frozen from this point. Under MW16 the contract administrator is not required to certify non-completion, but it may be prudent to record the failure in a letter to both the employer and the contractor. Once the date for completion has passed, the contractor is said to be in 'culpable delay'.

Liquidated and ascertained damages

4.44 The agreed rate for liquidated and ascertained damages is entered in the contract particulars (cl 2.8.1, or 2.9.1 in MWD16). This is normally expressed as a specific sum per week (or other period) of delay, to be allowed by the contractor in the event of failure to complete by the completion date. As a result of two decisions in the Supreme Court, it is no longer considered essential that the amount is calculated on the basis of a genuine pre-estimate of the loss likely to be suffered (*Cavendish Square Holdings* v *El Makdessi* and *ParkingEye Limited* v *Beavis*; see also *Alfred McAlpine Capital Projects* v *Tilebox*). Provided that the amount is not 'out of all proportion' to the likely losses, the damages will be recoverable without the need to prove the actual loss suffered, irrespective of whether the actual loss is significantly less or more than the recoverable sum (*BFI Group of Companies* v *DCB Integration Systems*). In other words, once the rate has been agreed, both parties are bound by it. Of course, for practical reasons, the rate should always be discussed with the employer before inclusion in the tender documents, and an amount that will provide adequate compensation included to cover, among other things, any additional professional fees that may be charged during this period. If 'nil' is inserted then this may preclude the employer from claiming any damages at all (*Temloc* v *Errill*), whereas if the clause is left blank then the employer may still be able to claim general damages.

Alfred McAlpine Capital Projects Ltd v *Tilebox Ltd* [2005] BLR 271

This case contains a useful summary of the law relating to the distinction between liquidated damages and penalties. A WCD98 contract contained a liquidated damages provision of £45,000 per week. On the facts, this was a genuine pre-estimate of loss and the actual loss suffered by the developer, Tilebox, was higher. The contractor therefore failed to obtain a declaration that the provision was a penalty. However, the judge also considered a different (hypothetical) interpretation of the facts whereby it was most unlikely, although just conceivable, that the total weekly loss would be as high as £45,000. In this situation also the judge considered that the provision would not constitute a penalty. In reaching this decision he took into account the fact that the amount of loss was difficult to predict, that the figure was a genuine attempt to estimate losses, that the figure was discussed at the time the contract was formed and that the parties at that time were represented by lawyers.

Cavendish Square Holdings v *El Makdessi* and *ParkingEye Limited* v *Beavis*, Supreme Court 2015

In this landmark case the Supreme Court restated the law regarding whether a liquidated damages clause may be considered a penalty. Key criteria for whether a provision will be penal are: if 'the sum stipulated for is extravagant and unconscionable in amount in comparison with the greatest loss that could conceivably be proved to have followed from the breach'; and whether the sum imposes a detriment on the contract breaker which is 'out of all proportion to any legitimate interest of the innocent party'. In determining these, the court must consider the wider commercial context.

BFI Group of Companies Ltd v *DCB Integration Systems Ltd* [1987] CILL 348

BFI employed DCB on the Agreement for Minor Building Works to refurbish and alter offices and workshops at its transport depot. BFI was given possession of the building on the extended date for completion, but two of the six vehicle bays could not be used for another six weeks as the roller shutters had not yet been installed. Disputes arose which were taken to arbitration. The arbitrator found that the delay in completing the two bays did not cause BFI any loss of revenue, and that BFI was therefore not entitled to any of the liquidated damages. BFI was given leave to appeal to the High Court. HH Judge John Davies QC found that BFI was entitled to liquidated damages. It was quite irrelevant to consider whether in fact there was any loss. Liquidated damages do not run until possession is given to the employer but until practical completion is achieved, which may not be at the same time. Therefore, the fact that the employer had use of the building was also not relevant.

Temloc Ltd v *Errill Properties* (1987) 39 BLR 30 (CA)

Temloc entered into a contract with Errill Properties to construct a development near Plymouth. The contract was on JCT80 and was in the value of £840,000. '£Nil' was entered in the appendix against clause 24.2, liquidated and ascertained damages. Practical completion was certified around six weeks later than the revised date for completion. Temloc brought a claim against Errill Properties for non-payment of some certified amounts and Errill counterclaimed for damages for late completion. It was held by the court that the effect of '£Nil' was not that the clause should be disregarded (because, for example, it indicated that it had not been possible to assess a rate in advance), but that it had been agreed that there should be no damages for late completion. Clause 24 is an exhaustive remedy and covers all losses normally attributable to a failure to complete on time. The defendant could not therefore fall back on the common law remedy of general damages for breach of contract.

4.45 The liquidated damages may either be recovered from the contractor as a debt or deducted from monies due (cl 2.8.2, or 2.9.2 in MW16). In both cases the following preconditions must have been met:

- the contractor must have failed to complete the works by the completion date (cl 2.8.1, or 2.9.1 in MWD16);
- the contract administrator must have fulfilled all duties with respect to the award of extensions of time.

4.46 If the employer wishes to deduct liquidated damages from an amount payable on a certificate, clause 2.8.2 (2.9.2 in MWD16) states that the employer must give a notice under clause 4.5.4 (a 'pay less notice'). The notice must be reasonably clear, and state the amount that the employer considers due, and the basis on which it is calculated, which means that information regarding the rate and period of damages should be included. In addition, if the employer wishes to deduct the damages from the sum due under the final certificate, this must have been made clear in writing before the date of the final certificate (cl 2.8.3, or 2.9.3 in MWD16). This is an earlier notification than that required under clause 4.5.4.

4.47 If an extension of time is given following the date for completion, the employer must immediately repay any liquidated damages recovered for the period up to the new completion date. In *Department of Environment for Northern Ireland* v *Farrans*, a case relating to JCT63, it was decided that the contractor has the right to interest on any repaid liquidated damages. This decision was criticised at the time (see the commentary in volume 19 of the Building Law Reports) and would be unlikely to be applied in relation to SBC16, where the wording is now different. However, as the wording in MW16 is similar to that in JCT63, the right to interest in relation to this form remains an open question.

Department of Environment for Northern Ireland v *Farrans* (Construction) Ltd (1982) 19 BLR 1 (NI)

Farrans was employed to build an office block under JCT63. The original date for completion was 24 May 1975, but this was subsequently extended to 3 November 1977. During the course of the contract, the architect issued four certificates of non-completion. By 18 July 1977 the employer had deducted £197,000 in liquidated damages but, following the second non-completion certificate, repaid £77,900 of those deductions. This process was repeated following the issue of the subsequent non-completion certificates. Farrans brought proceedings in the High Court of Justice in Northern Ireland, claiming interest on the sums that had been subsequently repaid. The court found for the contractor, stating that the employer had been in breach of contract in deducting monies on the basis of the first, second and third certificates, and that the contractor was entitled to interest as a result.

4.48 Certificates should always show the full amount due to the contractor. As explained in the guidance notes to the form, it is the employer alone that makes the deduction of liquidated damages. The employer should be advised on the completion date (as last revised) of the fact of failure to complete by the date and of the date of practical completion, and reminded of its right to deduct the damages and the procedure that must be followed. The employer would not be considered to have waived its claim by a failure to deduct damages from the first or any certificate under which this could validly be done, and would always be able to deduct the amount from a later certificate or to reclaim it as a debt at any point up until the final certificate.

5 Control of the works

5.1 The contract administrator derives authority solely from the wording of the contract and will, for example, supply necessary information, issue instructions and issue certificates or notices. In some matters the contract administrator will act as agent of the employer, for example when issuing instructions which vary the works, and in others will act as independent decision maker, such as when issuing certificates or deciding on claims for an extension of time. Failure to comply with any obligation (usually prefaced by the phrase 'the contract administrator shall') will constitute failure on the part of the employer tantamount to breach of contract (see Tables 5.1 and 5.2).

5.2 The day-to-day control of the works, i.e. the management of operations on site, co-ordination of orders and supplies, procurement of labour and sub-contractors and all issues relating to quality control and health and safety, is entirely the responsibility of the contractor.

Person in charge

5.3 The contractor is required to keep a competent 'person in charge' (cl 3.2) on the site at all reasonable times to receive any instructions given by the contract administrator and to act as the contractor's agent on site. Although there is no requirement in the contract conditions to have the person named, it would be sensible to set this requirement out in the specification, or alternatively to ask for the name and expected duration on site to be minuted at a pre-contract meeting. In extreme cases the contract administrator has the power to exclude persons from the works (cl 3.8), but this is only likely to be used where an employee may be seriously affecting operations on site.

Table 5.1 Key powers of the contract administrator

Clause (MW/MWD)	Contract administrator's express power
/2.5.2	Approve the contractor's proposal to deal with inconsistency in CDP documents
2.10/2.11	Instruct that defects can remain
3.3.1	Consent to the contractor sub-contracting the works or any contractor design
3.4/3.4.1	Issue written instructions to the contractor
3.6.1	Instruct variations, including additions to or omissions from the works and the order or period in which they are to be carried out
3.6.2	Agree the price of variations with the contractor before they are carried out
3.8	Exclude employed persons from the site
4.5.5	Issue a pay less notice on behalf of the employer
6.4.1	Give the contractor notice of defaults

Table 5.2 Key duties of the contract administrator

Clause (MW/MWD)	Contract administrator's duty
2.3/2.4	Issue any further information necessary
2.3/2.4	Issue all certificates
2.4/2.5.1	Correct inconsistencies between contract documents
2.7/2.8	Make such extensions of time as may be reasonable
2.9/2.10	Certify practical completion
2.10/2.11	Notify contractor of defects
2.11/2.12	Certify that defects have been made good
3.4/3.4.1	Confirm instructions in writing
3.6.2	Endeavour to agree value of variation with contractor
3.6.3	Value variation instructions
3.6.3	Ascertain amount of direct loss and/or expense
3.7	Issue instructions regarding expenditure of provisional sums
4.3	Issue interim certificates
4.8.2	Issue final certificate
5.6.5.1	Issue reinstatement certificates regarding insurance monies to be paid to the contractor
Supplemental Provision 1	Work collaboratively with other team members
Supplemental Provision 2	Establish a working environment where health and safety is of paramount concern
Supplemental Provision 3	Confirm cost-saving measure in an instruction

Clerk of works

5.4 There is no provision in MW16 for an independent clerk of works. If a clerk of works is to be engaged by the employer then this would have to be made clear at tender stage, and suitable provisions would have to be agreed regarding access and facilities.

Sub-contracted work

5.5 Under clause 3.3.1 the contractor may only sub-contract work with the written consent of the contract administrator. Failure to obtain this would be a default, although the contract sets out no remedy. Under clause 1.7.1, however, the contract administrator's permission cannot be unreasonably withheld. It is suggested that permission is required for each instance of sub-letting, rather than agreeing to sub-letting in principle.

5.6 MW16 stipulates that 'Where considered appropriate, the Contractor shall engage the sub-contractor using the JCT Short Form of Sub-Contract' (or the JCT Minor Works Sub-Contract with sub-contractor's design, in MWD16). This is not an absolute obligation, but it is suggested that the contractor would have to give reasonable justification for not using the form. Whatever form of domestic sub-contract is used, it must include certain conditions. The sub-contract must provide that the sub-contractor's employment must terminate immediately on termination of the contractor's employment (cl 3.3.2.1), and that the sub-contractor must comply with its CDM obligations (cl 3.3.2.2). Clause 3.3.2.3 states the sub-contract must provide that the sub-contractor has a right to interest on late payments by the contractor. It is surprising that the JCT decided to 'step down' such a lengthy and complex provision into the Minor Works form, when its sole purpose is to protect the interests of sub-contractors. It is notable that other SBC16 contract provisions, designed to protect the position of the employer in relation to unfixed goods and materials, have not been stepped down.

5.7 The contract makes no provision for naming or nominating a sub-contractor. It is suggested that if a sub-contractor is named in the tender documents, the contractor will remain entirely responsible for the performance of that sub-contractor. However, difficulties may arise if the sub-contractor were to repudiate its contract. The contract does not include a facility for replacement or substitution, and it may well be that the employer would be responsible for finding a replacement. Similarly, there may be problems if a design obligation is sub-contracted to a named specialist under MWD16 and defects emerge in relation to that design. Although by no means certain, it is possible that a court would find the main contractor not liable for the design error. It should also be remembered that there is no JCT employer/specialist warranty for use with MWD16 in the event that the employer wishes to be able to hold the specialist sub-contractor directly liable for this work.

5.8 It would be possible to limit the contractor's choice of a sub-contractor to carry out certain work to any one of three or more firms, in a manner similar to the provisions for naming firms in a list under SBC16. The National Building Specification MW version contains a suitable specification clause. Sometimes a contract administrator's instruction on the expenditure of a provisional sum is used as an opportunity to instruct the contractor to enter into a sub-contract with a particular firm. The contractor should be made aware of such an intention at the time of tendering, in order to reduce the risk of objections which could prove disruptive.

Work not forming part of the contract/persons engaged by the employer

5.9 There are no provisions whereby the employer may engage persons directly to carry out work that does not form part of the contract while the contractor is carrying out the works. While it is possible to make such arrangements, in practice it often results in significant problems as it is unclear who will be responsible for co-ordinating the work of the two contractors. If it is necessary, the specification would have to have set out this requirement, giving as much detail as possible about the nature and duration of the work. If the work should differ from the details set out, then this might be grounds for an extension of time and other claims by the contractor. It should be remembered that the indemnities given by the contractor under Section 5 of the contract (and therefore the insurances under Section 5) do not extend to persons not employed by the contractor.

Principal designer

5.10 Both parties are required to comply with the CDM Regulations (cl 3.9, see paragraph 3.21). It is the employer's obligation under the contract to ensure that the principal designer and the principal contractor carry out all the relevant duties under the CDM Regulations (cl 3.9.1). The contract administrator has no duty under MW16 to check that the contractor is complying with health and safety requirements on site (nor does the principal designer under the CDM Regulations), although the principal designer will become involved if design changes are needed, for example due to the discovery of an unexpected hazard. Generally, though, the responsibility for ensuring that correct health and safety measures are employed on site rests with the contractor.

Information to be provided by the contract administrator

5.11 MW16 accepts that the contract documents might not contain sufficient information to enable the project to be constructed. Even if the works have been fully specified, it is likely, for example, that information regarding assembly, location, detail dimensions, colours, etc. will be needed by the contractor during construction. Supply of this further information will usually form part of the contract administrator's duties to the employer under the terms of appointment.

5.12 MW16 refers to the contract administrator's obligation to provide 'any further information and instructions necessary for the proper carrying out of the Works' (cl 2.3). This obligation is repeated in the MWD16 version (cl 2.4), although it is suggested it does not extend to providing information relating to the CDP, except where there is an inconsistency or ambiguity in the employer's requirements. The forms do not require the information to be released under a contract administrator's instruction, but this is sound practice, as it would enable the clause 3.4 provisions to be brought into operation if necessary (see paragraph 5.19). If any of the information supplied introduces changes or additions to the works, it should certainly be covered by a contract administrator's instruction requiring a variation.

5.13 The contract sets out no time limits in regard to the provision of information. However, in order to avoid causing delay, it would be wise to recognise that information and instructions should be provided in sufficient time to allow the contractor to complete by the completion date or, if the contractor appears unlikely to complete by this date, at a date when it is reasonably necessary for the contractor to receive the information.

5.14 Under MW16 there is no requirement for the contractor to advise the contract administrator of when information may be needed. It might, therefore, be prudent to set up a procedure (possibly at site meetings) where the contractor is requested to list the information that may be needed and when.

Information provided by the contractor

5.15 The contractor as 'principal contractor' may be required by the principal designer to provide information in relation to the health and safety file (CDM Regulations, regulation 12(7)). It should be noted, however, that MW16 does not contain express provisions for 'as-built' drawings. If these are needed, the specific requirement should be set out in the specification or schedules.

5.16 MWD16 includes limited provisions regarding the supply of information relating to the CDP. The contractor is required to provide the contract administrator with copies of 'such drawings or details, specifications of materials, goods and workmanship, and (if requested) related calculations and information, as are reasonably necessary to explain the Contractor's Designed Portion' (cl 2.1.3). The contractor may not commence the related work until after seven days from the date the information is supplied (cl 2.1). In practice, seven days is little time to consider all the implications of integrating the CDP with the rest of the project, and the administrator will need to act swiftly if the information reveals any potential problems. There are no provisions to deal with any comments the contract administrator might wish to make on the information. Although this does not prevent the contract administrator from commenting, the contractor is not obliged to incorporate the comments. If an agreement cannot be reached on matters raised, the contract administrator may need to instruct a variation to the CDP under clause 3.6.1 (see paragraph 5.19).

Inspection and tests

5.17 On most projects the contract administrator will inspect the works at regular intervals. MW16 does not place a duty on the contract administrator to do this, however the contract administrator's obligations to the employer will, almost always, include a duty to inspect. Normally this would be an express duty under the terms of appointment, but in some circumstances it could also be implied: clearly, when the contract administrator is required under the contract to form an opinion on various matters – including being satisfied with the standard of work and materials prior to issuing a payment certificate – then it is essential that some form of inspection takes place. However, it is important to note that the duty is owed to the employer, and not to the contractor. For example, a contractor cannot blame a contract administrator for failing to draw its attention to defective work.

5.18 Furthermore, a contract administrator will not necessarily be liable to the employer for negligent inspection if a defect in a contractor's work is not identified. The question in every case is whether the contract administrator exhibited the degree of skill that an ordinary competent professional would exhibit in the same circumstances. Generally, the extent and frequency of inspections must enable the contract administrator to be in a position to properly certify that the construction work has been carried out in accordance with the contract (*Jameson* v *Simon*). The case of *McGlinn* v *Waltham Contractors Ltd* sets out some useful advice on the appropriate standard of inspection.

McGlinn v *Waltham Contractors Ltd* (2007) 111 Con LR 1

This case concerned a house in Jersey called *Maison d'Or* that was designed and built for the claimant, Mr McGlinn. The house took three years to build, but after it was substantially complete, it sat empty for the next three years while defects were investigated. It was completely demolished in 2005 having never been lived in and was not rebuilt. Mr McGlinn brought an action against the various consultants, including the architect, and the contractor, claiming that *Maison d'Or* was so badly designed, and so badly built, that he was entitled to demolish it and start again. The contractor however had gone into administration and played no part in the hearing. The architect was engaged on RIBA Standard Form of Appointment 1982, which referred to 'periodic inspections'. HH Judge Peter Coulson QC usefully summarised the principles relating to inspection (at paras 215 and 218), which included the following:

- The change from 'supervision' to 'inspection' represented 'a potentially important reduction in the scope of an architect's services'.

- 'The frequency and duration of inspections should be tailored to the nature of the works going on at site from time to time'.
- 'If the element of the work is important because it is going to be repeated throughout one significant part of the building, then the inspecting professional should ensure that he has seen that element of the work in the early course of construction/assembly so as to form a view as to the contractor's ability to carry out that particular task'.

Contract administrator's instructions

5.19 The contract administrator has the power to issue instructions (cl 3.4, or 3.4.1 in MWD16). Sometimes the contract states that the contract administrator may issue instructions (e.g. instructions requiring a variation under clause 3.6.1), but at other times the contract states that the contract administrator shall issue instructions (e.g. instructions regarding provisional sums under clause 3.7). The latter is a contractual obligation and failure by the contract administrator to issue the instruction will constitute a breach of contract by the employer. If the employer gives an instruction other than through the contract administrator, this would be of no effect under the contract and the contractor would be under no obligation to comply with any such instruction. If the contractor does, however, carry out the instruction, a court might consider that there had been an agreed amendment to the contract, but the consequences would be difficult to sort out in practice and the employer would be very unwise to risk such action.

5.20 Instructions empowered by the contract are:

- acceptance of defective work (cl 2.10/2.11);
- changes in the CDP or the works (variations) (cl 3.6.1);
- expenditure of provisional sums (cl 3.7);
- exclusion of persons from the works (cl 3.8).

5.21 Clause 3.4 (3.4.1 in MWD16) requires the contractor to comply with written instructions 'forthwith'. If any instructions are given orally, the contract administrator must confirm the instruction in writing, otherwise it would be of no effect. It would therefore be sensible to confirm all instructions immediately. Although MW16 makes no reference to the contractor confirming oral instructions in writing, the contract administrator would be wise to check any such confirmation and respond. If the contract administrator remains silent, it may be deemed that the contract administrator has approved the contractor's version. If the contractor carries out work on the basis of an oral instruction only, then it is suggested that the contract administrator could later sanction the instruction at any time prior to the issue of the final certificate, but the contractor would be taking a risk.

5.22 There is no special format required for instructions, but it is often convenient to use the forms published by RIBA Publishing or NBS Contract Administrator. Instructions in site meeting minutes might constitute a written confirmation of an oral instruction if issued by the contract administrator, but are unlikely to do so if issued by the contractor. It would depend on the circumstances whether the minutes were sufficiently clear to fall within the terms of the contract and it is therefore not good practice to rely on this method.

5.23 The contractor must comply with every instruction, provided that it is valid and one which the contract administrator is empowered to issue. The contractor must 'forthwith' comply, which for practical purposes means as soon as is reasonably possible (cl 3.4, or 3.4.1 in MWD16).

5.24 If the contractor does not comply with a written instruction, the employer may employ and pay others to carry out the work to the extent necessary to give effect to the instruction (cl 3.5). The contract administrator must already have given notice to the contractor requiring compliance with the instruction, and seven days must have elapsed after the contractor's receipt of the notice before the employer may bring in others. This suggests that some recorded form of delivery is desirable. Careful records should be kept to substantiate the costs claimed and competitive tenders obtained if the circumstances permit. If the contractor believes that a contract administrator's instruction might not be empowered by the contract, or justifies clarification, then its last resort would be to raise the matter in adjudication. Any additional costs to the employer, i.e. the difference between what would have been paid to the contractor for the instructed work and the costs actually incurred by the employer, are to be deducted from the contract sum. These costs could include not only the carrying out of the instructed work but also any special site provisions that would need to be made, including health and safety provisions, and any additional professional fees charged. Although it would be wise to obtain alternative estimates for all these costs wherever possible, if the work is needed urgently there would be no need to do so.

Variations

5.25 The contract administrator's instructions often require variations to the works. Under common law neither party to a contract has the power to unilaterally alter any of its terms. Therefore, neither the employer nor the contract administrator would have the power to require any variations unless the contract provides for this. As it is difficult to define some things exactly in advance, most construction contracts contain provisions allowing the employer to vary the works to some degree. Changes can arise because of unexpected site problems, or because of design changes wanted by the employer, or because the contract administrator has to change information issued to the contractor.

5.26 Under MW16 the contract administrator has the power to order variations, and the scope of what constitutes a variation is set out in clause 3.6.1. It includes 'an addition to, omission from, or other change in the Works or the order or manner in which they are to be carried out'. The power does not extend to altering the nature of the contract, nor can the contract administrator issue variations after practical completion. In addition, the word 'manner' replaces the word 'period' used in the 2011 edition; this change makes it clear that an instruction cannot remove work and require it to be carried out at another time, for example during the rectification period. All variations under clause 3.6.1 may result in an adjustment of the contract sum and give rise to a claim for an extension of time and direct loss and/or expense. If the works are suspended as a consequence, this could also be a ground for determination by the contractor.

5.27 The contract administrator may vary the works (e.g. by changing the standard of a material specified), and may add to or omit work, or substitute one type of work for another. Although there is no specific provision in MW16, it seems likely that the contract administrator could order removal of work already carried out. The contract administrator is also empowered to order variations affecting the sequence of work. Although not entirely clear, it is likely that the phrase 'manner of carrying out the work' could be interpreted broadly, to include not just working methods, but also access to or use of the site, limitations on working space or working hours or any restrictions already imposed. In any case, the contractor is unlikely to challenge such an instruction, given that the contractor will be paid for such variations and would be entitled to an extension of time. It would also seem likely that a court would imply such a term, at least in relation to restrictions that might be imposed by a local authority, as otherwise the contract might become unworkable.

5.28 Under MWD16 the contract administrator may also instruct a change to the employer's requirements which necessitates an alteration to the design of the CDP works. This provision is carefully worded, and does not empower the administrator to instruct changes to the CDP directly. Under clause 2.1.4, the contractor will not be responsible for any changes to the requirements. In effect, although the contractor is responsible for the design of the CDP, the employer may not be able to hold the contractor responsible for problems resulting from such variations.

Defective work

5.29 Clause 2.1 states that all materials and workmanship shall be of the standard specified in the contract documents. The contract administrator will normally inspect at regular intervals to monitor the standard that is being achieved. If any changes were made in order to raise or lower the standard then this would constitute a variation. When the standard achieved appears to be unsatisfactory it can be all too easy for the contract administrator to become involved in directing the day-to-day activities of the contractor on site, particularly with smaller jobs under a contract such as MW16. Apart from being an enormous burden on the contract administrator, this could confuse the issue of who is ultimately responsible for the works and must be avoided at all costs. The contract administrator would normally, of course, draw the contractor's attention to areas of defective or poor-quality work.

5.30 Clearly the contractor should not be paid for any defective work. If this appears to be insufficient incentive then the contract administrator may instruct that the work in question is carried out in accordance with the contract. While not adding anything to the contractor's primary obligation, this allows the employer to employ others to carry out the work should the contractor refuse to comply, provided notice is given (cl 3.5). If the contractor insists that the work was correctly carried out then this dispute might have to be taken to adjudication. If the contract administrator wishes to have tests carried out then the cost of these would be borne by the employer, unless special provisions have been set out in the contract documents, as MW16 has no provision for testing or opening up work.

5.31 The contract does not expressly give the contract administrator the power to accept defective work or instruct that it remains (except during the rectification period, see paragraph 5.34). However, the same effect could be achieved by issuing a variation instruction that effectively lowers the standard of work set out in the contract documents to that which has actually been provided. The contract administrator should not do this without the employer's consent, which should be recorded in writing. It would be inadvisable unless the employer is completely satisfied with the work, the defect is minor and the delay consequent upon correcting the work appears completely out of proportion to the benefit to be gained. It is very important that the parties agree on any deduction to be made before the defective work is accepted (see paragraph 7.8). The contract administrator should strongly advise the employer against accepting any defective work that could later cause technical problems or be a source of irritation. The difficult case of *Ruxley Electronics* v *Forsyth* illustrates that it may not be possible to claim the cost of having the work rebuilt at a later date.

Ruxley Electronics and Construction Ltd v *Forsyth* (1995) 73 BLR 1 (HL)

Mr Forsyth employed Ruxley Electronics to build a swimming pool. The drawings and specification required the pool to be 7ft 6in. deep at its deepest point, but the completed pool was only 6ft 9in. deep. The contractor brought a claim for its unpaid account, and Mr Forsyth counterclaimed the cost of rebuilding the pool, which would be £21,560. The trial judge found that the shortfall in depth did not decrease the value of the pool and that Mr Forsyth had no intention of building a new pool. He rejected the counterclaim but awarded £2,500 as general damages for loss of pleasure and amenity. Mr Forsyth appealed and the Court of Appeal allowed the appeal and awarded him £21,500. The contractor appealed and the House of Lords restored the original ruling, confirming that the cost of reinstatement is not the only possible measure of damages for defective performance of a building contract and is not the appropriate measure where the expenditure would be out of all proportion to the benefit to be obtained.

Making good defects

5.32 The contractor is required to make good any 'defects, shrinkages or other faults' to the works which appear during the rectification period, unless the contract administrator instructs otherwise (cl 2.10, or 2.11 in MWD16). This is stated in the contract particulars to be three months, although a different period can be inserted if required. A longer period may be advisable, particularly where there are mechanical services which need to be tested over a range of outdoor temperatures. The obligation extends to defects resulting from a failure of the contractor to comply with its CDM obligations, but would not include defects that may be due, for example, to errors in the design information supplied to the contractor or to general wear and tear resulting from occupation by the employer. It is suggested also that although this contract provision is limited to those defects that appear after practical completion, and does not extend to defects that were patent at that time, in practice it would be sensible to allow the contractor the opportunity to remedy any such defects (*William Tomkinson* v *Parochial Church Council of St Michael*). The obligation to return to site is, in fact, of benefit to the contractor as it carries with it a corollary right to have access to the site to make good its own defaults. If the clause were not present, the employer would have the right to employ another firm and bring a claim for damages against the contractor. The cost to the contractor would almost certainly be greater than carrying out the work itself. Although the right to return to site ceases at the end of the three-month period, the contractor's liability for defective materials or workmanship continues throughout the statutory limitation period.

William Tomkinson & Sons Ltd v *Parochial Church Council of St Michael* (1990) 6 CLJ 319, 8 CLD-08-05

The Council employed William Tomkinson to carry out repair works to their church caused by the contractor's negligence, on the Agreement for Minor Building Works. During the course of the works £100,000 worth of damage was caused to the church organ and other parts of the structure by rainwater. The architect reported complaints about these defects to the contractor orally, but did not issue a schedule of defects. The Council employed other contractors to remedy the defects prior to the date for practical completion. In examining clause 2.5, Judge Stannard concluded that a written schedule of defects was not necessary and that oral notification was sufficient. He also found that the words 'appear within three months of practical completion' extended to defects which appear before practical completion. The true measure of damages was not the church's outlay in repairing the damage but what it would have cost the contractor if it had been required to undertake the repairs.

5.33 MW16 requires the contract administrator to notify the contractor of the existence of defects within 14 days of the end of the rectification period (cl 2.10, or cl 2.11 in MWD16). Unlike SBC16, the form does not state that the contract administrator must issue a schedule of defects. It is suggested that it would be sufficient for the contract administrator to write to the contractor to inform it that defects had appeared, and of their general nature. The onus would then be on the contractor to identify and make good all defective work. If the contract administrator prefers to issue a schedule, it might be wise to state that it is not intended to be a comprehensive list, and that the contractor should make its own inspection. It would also be sensible to make the employer aware that the contractor must be allowed access, as to prevent this might result in the employer being unable to claim for the costs of remedying the defective work (see *Pearce and High* v *Baxter*).

Pearce and High Ltd v *John P Baxter and Mrs A S Baxter* [1999] BLR 101 (CA)

The Baxters employed Pearce and High on MW80 to carry out certain works at their home in Farringdon. Following practical completion, the architect issued interim certificate no. 5, which the employer did not pay. The contractor commenced proceedings in Oxford County Court, claiming payment of that certificate and additional sums. The employer in its defence and counterclaim relied on various defects in the work that had been carried out. Although the defects liability period had by that time expired, neither the architect nor the employer had notified the contractor of the defects. The Recorder held that clause 2.5 was a condition precedent to the recovery of damages by the employer, and further stated that it was a condition precedent that the building owner had notified the contractor of patent defects within the defects liability period. The employer appealed and the appeal was allowed. Lord Justice Evans stated that there were no clear express provisions within the contract which prevented the employer bringing a claim for defective work, regardless of whether notification had been given. He went on to state, however, that the contractor would not be liable for the full cost to the employer of remedying the defects, if the contractor had been effectively denied the right to return and remedy the defects itself.

5.34 If the employer would prefer to accept the defects rather than require them to be corrected, then an appropriate deduction is made from the contract sum (cl 2.11). As noted above (see paragraph 5.31), care should be taken to establish the full extent of the problem and negotiate a deduction before such a course of action is taken as it is unlikely that the employer would thereafter be able to claim for consequential problems or further remedial work.

5.35 Once satisfied that the contractor's obligations have been discharged, the contract administrator must issue a certificate to that effect (cl 2.11, or 2.12 in MWD16). The certificate is a precondition to the issue of the final certificate. The contract does not state what should happen in respect of defects that appear after the issue of the certificate but before the issue of the final certificate. It appears, however, that the contractor may no longer have any obligation under clause 2.10 (2.11 in MWD16) nor any right under the contract, to return to site. It is suggested that in such circumstances there would be two possible courses of action. The first would be to make an agreement with the contractor to rectify the defects before the final certificate is issued. If the contractor were to refuse to do this, an amount could be deducted from the contract sum to cover the cost of making good the work, but this might involve some risk to the employer. The second and less risky course would be to have the defective work rectified by another contractor, and to deduct the amount paid from the contract sum. This would involve a delay to the issue of the final certificate and would probably be disputed by the contractor.

6 Sums properly due

6.1 The contract sum is entered in Article 2. This is seldom the amount which the project will ultimately cost, and the wording of the contract recognises this by the qualifying reference 'or such other sum as becomes payable'.

6.2 The contract figure sometimes contains provisional sums to cover the cost of work that cannot be accurately described or measured at the time of tendering. Also, almost all jobs will entail some variations as work proceeds; MW16 provides for dealing with the cost of such variations, for the 'direct loss and/or expense' due to any resulting disruption, and for costs and expenses due to suspension. If Supplemental Provision 3 is included, the contractor may propose savings that result in an adjustment to the contract sum. The form can take into account fluctuations – limited to 'contribution, levy and tax changes' – or, if preferred, it can be operated as literally a fixed price contract. VAT is, of course, not included in the contract sum.

6.3 As a result, the contract sum will almost certainly need to be adjusted, and amounts properly ascertained by the contract administrator will be added or deducted as appropriate when payments are certified.

6.4 Arithmetical errors by the contractor in pricing are not allowed as a cause for adjustment. The contractor has agreed to carry out all the work shown in the contract documents for the contract sum, and errors or omissions in any detailed pricing breakdown are immaterial. Any inconsistency between the contract drawings and other documents must be corrected by a contract administrator's instruction and is treated as if it were a variation, but it is suggested that this would not necessarily result in an increase in the contract sum (cl 2.4, or 2.5.1 in MWD16). If the work is clearly shown in some of the documents but does not appear in others, a court would be likely to ask whether, from the point of view of an objective bystander, it is clear that the parties intended that this work should form part of their agreement. The contract administrator should make a similar objective assessment, and if the answer is 'no' then the variation will result in an adjustment.

Provisional sums

6.5 If sufficient information cannot be provided at the time of tender to allow the contractor to price that item then a provisional sum may be included in the tender documents to cover the item. Provisional sums included in the contract specification are not subject to the New Rules of Measurement (NRM2 Rules) (RICS, 2012). It is nevertheless advisable to give the tenderer as much information as possible regarding the nature and construction of the work, how and where the work fits into the building, the scope and extent of the work and any specific limitations on methods or sequence or timing. The contract administrator must issue instructions regarding all work covered by provisional sums (cl 3.7). The work is then to be valued in accordance with clause 3.6.2 or 3.6.3. It should be noted that a provisional sum is different to a 'prime cost' (PC) sum. The latter normally refers to a price that has been ascertained in advance of requesting tenders, from a specialist supplier or sub-contractor. It is unlikely that PC sums will be applicable to a contract based on MW16, as the form does not make provision for the naming or nominating of such firms (see paragraph 1.16).

Valuation of variations

6.6 Clause 3.6.3 makes it clear that valuation of variations is the responsibility of the contract administrator, regardless of whether or not a quantity surveyor is appointed. The clause applies to variations instructions and to 'matters to be treated as a variation', which includes correction of inconsistencies under clause 2.4 (2.5.1 in MWD16) and reinstatement work following damage under clause 5.6.6. Work is to be valued 'using any relevant prices' in the priced specification or schedules or in the contractor's schedule of rates. 'Relevant' should be understood in the sense that the variation relates to work of a similar character, carried out under similar conditions. Where the work is not of similar character it should be valued on a fair and reasonable basis. Any direct loss and/or expense incurred is to be taken into account in the valuation (see paragraph 6.8).

6.7 Under clause 3.6.2 the contract administrator should endeavour to agree a price with the contractor in advance of the work being carried out. The contract administrator would, however, be acting as agent for the employer, and it would be wise to confirm the price with the employer before agreeing to it. The price could be agreed using the Supplemental Provision 3 procedure. Alternatively, a method similar to the 'Schedule 2 quotation' procedure in SBC16 could be adopted, whereby a quotation is requested which should identify the direct cost of complying with the instruction, the period required for extension to the contract period and any amount necessary to cover direct loss and/or expense. It may not be generally appropriate for contracts under MW16 because of the sums involved and the time factor, but for more significant variations this method would bring certainty of outcome for the parties, as both would be bound by what is agreed. The certainty, however, is likely to be secured only at a price, as the contractor would be under no obligation to relate the quotation to the figures in the contract documents. If the quotation is unacceptable, the work could still be instructed and valued in the usual way.

Direct loss and/or expense

6.8 Clause 3.6.3 requires that valuations of variations, and all matters to be treated as a variation, shall include the amount of any 'direct loss and/or expense'. This is the only instance within MW16 where the contract administrator may make such an award. If the contractor suffers losses not related to a variation, these might have to be referred to adjudication, arbitration or litigation, unless some agreement can be reached. The phrase 'direct loss and/or expense' refers to losses suffered as a result of delay or disruption consequent upon the variation, excluding, of course, the direct cost of carrying out the relevant work. The clause also refers to loss and/or expense incurred due to compliance or non-compliance by the employer with health and safety obligations as set out in clause 3.9. This relates purely to variations, and in theory it is possible to envisage situations where losses are caused to the contractor through the employer's compliance with the CDM Regulations, which do not, strictly speaking, constitute a variation to the works. Such losses would have to be dealt with outside the terms of the contract.

6.9 In ascertaining loss and expense the contract administrator must determine what has actually been suffered. The sums awarded can include any loss or expense that has arisen directly as the result of the variation. In assessing the amount of damages the object is to put the contractor back into the position in which it would have been but for the disturbance. The contractor ought to be able to show that it has taken reasonable steps to mitigate its loss.

6.10 The following are items which could be included:

- increased preliminaries;
- overheads;
- loss of profit;
- uneconomic working;
- increases due to inflation;
- interest or finance charges.

6.11 Prolongation costs, such as on-site overheads, would normally only be claimable for variations that result in an extension of time. (For head office overheads, etc., see *Alfred McAlpine* v *Property and Land Contractors*.) Interest may also be recoverable, but only if it can be proved to have been a genuine loss (*F G Minter* v *WHTSO*).

Alfred McAlpine Homes North Ltd v *Property and Land Contractors Ltd* (1995) 76 BLR 59

An appeal arose on a question of law arising out of an arbitrator's award regarding the basis for awarding direct loss and expense with respect to additional overheads and hire of small plant, following an instruction to postpone the works. The judgment contains useful guidance on the basis for awarding direct loss and expense. To 'ascertain' means to 'find out for certain'. It is not necessary to differentiate between 'loss' and 'expense' in a head of claim. Regarding overheads, a contractor would normally be entitled to recover as a 'loss' the shortfall in the contribution that the volume of work had been expected to make to the fixed head office overheads, but which, because of a reduction in volume and revenue caused by the prolongation, was not in fact made. The fact that 'Emden' or 'Hudson' formulae depend on certain assumptions means that they are frequently inappropriate. The losses on the plant should be the true cost to the contractor, not based on notional or assumed hire charges.

F G Minter Ltd v *Welsh Health Technical Services Organisation* (1980) 13 BLR 1 (CA)

Minter was employed by Welsh Health Technical Services Organisation (WHTSO) under JCT63 to construct the University Hospital of Wales (second phase) Teaching Hospital. During the course of the contract several variations were made, and the progress of the works was affected due to the lack of necessary drawings and information. The contractor was paid amounts in respect of direct loss and/or expense, but the amounts paid were challenged as insufficient. The amounts had not been certified and paid until long after the losses had been incurred, therefore the amounts should have included an allowance in respect of finance charges or interest. Following arbitration, several questions were put to the High Court, including whether Minter was entitled to finance charges in respect of any of the following periods:

(a) between the loss and/or expense being incurred and the making of a written application for the same;
(b) during the ascertainment of the amount; and/or
(c) between the time of such ascertainment and the issue of the certificate including the ascertained amount.

The court answered 'no' to all three questions and Minter appealed. The Court of Appeal ruled that the answer was 'yes' to the first question and 'no' to the others.

6.12 Although the contract does not actually state such a requirement, it would be reasonable to assume that the contractor should provide details and particulars of all items concerned with any loss or expense, otherwise it will be extremely difficult for the contract administrator to assess what amount would be reasonable. If no information is forthcoming, the contract administrator should nevertheless try to make a fair and reasonable assessment.

6.13 Formulae such as the 'Hudson' or 'Emden' formulae are sometimes used to estimate head office overheads and profit, which may be difficult to substantiate. Such formulae can only be used where it has been established that there has been a loss of this nature. To do this the contractor must be able to show that, but for the delay, the contractor would have been able to earn the amounts claimed on another contract, for example by producing evidence such as invitations to tender which were declined. Such formulae may be useful where it is difficult to quantify the amount of the alleged loss, provided a check is made that the assumptions on which the formula is based apply.

6.14 Although direct loss and/or expense is a matter of money and not time, which are quite separate issues, there is often a practical correlation in the case of prolongation. However, any general implication that there is a link would be incorrect and, in principle, disruption claims and delay to progress are independent. An extension of time, for example, is not a condition precedent to the award of direct loss and/or expense (*H Fairweather & Co.* v *Wandsworth*).

H Fairweather & Co. Ltd v *London Borough of Wandsworth* (1987) 39 BLR 106

Fairweather entered into a contract with the London Borough of Wandsworth to erect 478 dwellings. The contract was on JCT63. Pipe Conduits Ltd was nominated sub-contractor for underground heating works. Disputes arose and an arbitrator was appointed who made an interim award. Leave to appeal was given on several questions of law arising out of the award. The arbitrator had found that where a delay occurred which could be ascribed to more than one event, the extension should be granted for the dominant reason. Strikes were the dominant reason, and the arbitrator had granted an extension of 81 weeks for this reason, and made it clear that this reason did not carry any right for direct loss and/or expense. The court stated that an extension of time was not a condition precedent to an award of direct loss and/or expense, and that the contractor would be entitled to direct loss and/or expense for other events which had contributed to the delay.

6.15 Applications by or claims from the contractor must be dealt with according to the procedures contained in the contract. Failure to certify an amount properly due will not prevent recovery, and could leave the employer liable in damages for breach of contract (*Croudace* v *London Borough of Lambeth*).

***Croudace Ltd* v *The London Borough of Lambeth* (1986) 33 BLR 20 (CA)**

Croudace entered into an agreement with the London Borough of Lambeth to erect 148 dwelling houses, some shops and a hall. The contract was on JCT63 and the architect was Lambeth's chief architect and the quantity surveyor was its chief quantity surveyor. The architect delegated his duties to a private firm of architects. Croudace alleged that there had been delays and that they had suffered direct loss and/or expense and sent letters detailing the matters to the architects. In reply, the architects told Croudace that they had been instructed by Lambeth that all payments relating to 'loss and expense' had to be approved by the Borough. The chief architect of Lambeth then retired and was not immediately replaced. There were considerable delays pending the appointment and Croudace began legal proceedings. The High Court found that Lambeth was in breach of contract in failing to take the necessary steps to ensure that the claim was dealt with, and was liable to Croudace for this breach. The Court of Appeal upheld this finding.

Fluctuations

6.16 Depending upon the economic climate it may be to the employer's advantage to ask for a 'fixed' or 'guaranteed' price. Rather confusingly, the term 'fixed price' in building contracts usually includes for limited fluctuations, such as tax changes. MW16, however, can be operated literally as a fixed price contract (i.e. the contract sum will not change unless variations occur or provisional sums are included).

6.17 MW16 allows 'contribution, levy and tax changes' fluctuations, i.e. those which result from the intervention of statute after the 'Base Date' entered in the contract particulars. The details of how this operates are set out in Schedule 2. There is an opportunity for entering a 'percentage addition', which is unlikely to be in excess of 10 per cent. These limited fluctuation provisions are optional, but would operate unless otherwise indicated in the contract particulars. The contract particulars allow the parties to select 'no fluctuations' or to stipulate their own fluctuations provisions. If 'no fluctuations' is chosen, clause 4.9 operates to make the contract truly 'fixed price'; this would be appropriate with contracts of limited duration, which in practice would be any contract of less than 12 months.

6.18 Where a contract includes for fluctuations then, in the absence of anything to the contrary, these will be payable for the whole time the contractor is on site, even if it fails to complete within the contract period (*Peak Construction* v *McKinney Foundations Ltd*). There is a

***Peak Construction (Liverpool) Ltd* v *McKinney Foundations Ltd* (1970) 1 BLR 111 (CA)**

Peak Construction Ltd was main contractor on a contract to construct a multi-storey block of flats for Liverpool Corporation. As a result of defective work by nominated sub-contractor McKinney Foundations, work on the main contract was halted for 58 weeks, and the main contractor brought a claim against the sub-contractor for damages. The Official Referee, at first instance, found that the entire 58 weeks was delay caused by the nominated sub-contractor, and awarded £40,000 of damages, £10,000 of which was for rises in wage rates during the period. McKinney appealed, and the Court of Appeal found that the award of £10,000 could not be upheld as clause 27 of the main contract entitled Peak Construction to claim this from Liverpool Corporation right up until the time when the work was halted.

so-called 'freezing' provision in MW16 (Schedule 2:10.1), but this depends on the text of clause 2.7 ('Extensions of time'; cl 2.8 in MWD16) being left unamended and all written notices under clause 2.7 (or 2.8) having been properly dealt with by the contract administrator (Schedule 2:10.2.2).

7 Certification

7.1 One of the most important duties of the contract administrator under MW16 is to certify sums properly due to the contractor. Whether or not a quantity surveyor is employed, the contract administrator is responsible for both the valuations and for issuing certificates. It is a duty to be exercised with care and skill, and failure in this respect could amount to negligence. On the one hand, the contractor depends on the cash flow which proper payment should provide, and is entitled to be paid in accordance with the terms of the contract. On the other hand, overvaluation and certification could place the employer's interests at risk should the contractor become bankrupt, perhaps leaving a legacy of faulty workmanship and disputes over the ownership of unfixed materials. The contract administrator's duty to the employer has been fairly clear since the case of *Sutcliffe* v *Thackrah*, and it seems possible that there is a duty also to the contractor following *Salliss & Co.* v *Calil and W F Newman & Associates*. This is a difficult area of the law which is constantly developing, and it would be sensible for the contract administrator to proceed on the basis that such a duty of care exists.

Sutcliffe v *Thackrah* (1974) 4 BLR 16 (CA)

An architect issued certificates on a contract for the construction of a dwelling house. The contractor's employment was determined for proper reasons, following which the contractor was declared bankrupt. It then became apparent that much of the work, which had been included in the interim certificates, was defective, and the architect was found negligent. In the House of Lords, when reviewing the role of the architect, Lord Reid stated (at page 21):

> Many matters may arise in the course of the execution of a building contract where a decision has to be made which will affect the amount of money which the contractor gets. ... The building owner and the contractor make their contract on the understanding that in all such matters the architect will act in a fair and unbiased manner and it must therefore be implicit in the owner's contract with the architect that he shall not only exercise due care and skill but also reach such decisions fairly, holding the balance between his client and the contractor.

Michael Salliss & Co. Ltd v *Calil and William F Newman & Associates* (1987) 13 Con LR 69

Calil employed contractor Michael Salliss for some refurbishment works on JCT63. W F Newman acted as architect and quantity surveyor under the contract. The contractor commenced proceedings against the employer and joined the architect as second defendants, claiming that the architect was in breach of his duty to use all professional skill and care in granting only a 12-week extension of time when a 29-week extension was due.

There was a sub-trial as to whether the contractor could recover damages against the architect. HH Judge Fox-Andrews held that under a JCT contract the architect owed a duty to the contractor to act fairly between the employer and contractor in matters such as certification and extensions of time. He also noted (at page 79) that:

> in many respects an architect in circumstances such as these owes no duty to the contractors. He owes no duty to contractors in respect of the preparation of plans and specifications or in deciding matters such as whether or not he should cause a survey to be carried out. He owes no duty of care to a contractor whether or not he should order a variation. Once, however, he has ordered a variation he has to act fairly in pricing it.

Although this case was followed by another where an engineer was found to have no duty to the contractor (*Pacific Associates Inc.* v *Baxter*), in that instance the contract contained a special provision purporting to exempt the engineer from liability.

7.2 The issue of payment certificates is referred to in clauses 4.3 and 4.8, but there is little on the procedures to be adopted. It would be sensible to issue the certificate to the employer with a copy to the contractor at the same time. The procedure to be used should be established at the outset, either by setting it out in the contract documents or agreeing it at a pre-contract meeting.

Interim payments

7.3 Payments are to be made by the employer to the contractor after the issue of certificates by the contract administrator (cl 4.3). These are to be issued within five days of the 'due date', which in turn relates to the 'Interim Valuation Date' (see Figure 7.1). The interim valuation dates are entered into the contract particulars; the first should be no later than one month after the works commencement date, and subsequent dates at no greater than monthly intervals. If nothing is inserted, the first date is one month after the commencement date, and subsequent due dates occur at intervals of one month up until the final payment.

7.4 Unlike some other JCT forms, MW16 makes no provision for advance payment to the main contractor, nor for payment on commencement of the work (in effect the same thing), so if this is required, the contract terms would have to be amended. There is always a risk in making any such payment, and the contract administrator and employer should be quite clear as to what compensatory benefits, such as a reduction in the contract sum, would result.

7.5 Under clause 4.3 it is the contract administrator's responsibility to determine the value of the interim payment, even though the task might be delegated to a quantity surveyor. Clause 4.4.1 gives the contractor the right to make an application for payment, stating the amount it considers due, but although this information may be very useful the contract administrator must nevertheless make an independent assessment. The clause requires that certificates state not only the amount to be paid, but also 'the basis on which that sum has been calculated, including the amount of each adjustment'. It is unlikely that a great deal of detail will be required here; a short schedule will probably be sufficient, but it should include the various elements of the calculation as detailed under clause 4.3. Similar provisions are included for the final certificate.

7.6 Clause 4.3 states that interim payments should include the value of the work properly executed and the value of materials and goods properly brought onto site. The value of the work will be calculated using the prices and rates shown in the specification or schedules, or the contractor's schedule of rates, whichever is appropriate. There is no provision for the value of off-site materials, goods or prefabricated items. The valuation should, of course, take into account any relevant variations, including any instructions issued under clause 5.4B.2 for reinstatement after fire, etc. It should also take into account adjustments due to instructions relating to provisional sums (cl 3.7), and any costs and expenses due to suspension should also be added (cl 4.7). The total is then adjusted in accordance with the fluctuations provisions, if operated (Schedule 2). From this the following should be deducted: the total amount stated as due under previous certificates, any amount paid as a consequence of a contractor's payment notice issued since the last certificate, and any amounts deductible due to acceptance of defective work (cl 2.10, or cl 2.11 in MWD16) or to non-compliance with an instruction (cl 3.5). The total amount is then subject to a percentage reduction (see paragraph 7.13).

Value of work properly executed

7.7 The contract administrator should only certify after having carried out an inspection to a reasonably diligent standard. The contract does not set a date, but normally the certificate should include for work carried out up to seven days before the date of the certificate. Contract administrators should not include any work that appears not to have been properly executed, whether or not it is about to be remedied or the retention is adequate to cover remedial work (*Townsend* v *Stone Toms* and *Sutcliffe* v *Chippendale & Edmondson*). If a quantity surveyor has been appointed by the client, the contract administrator should also note the case of *Dhamija* v *Sunningdale Joineries Ltd*, which stated that a quantity surveyor is not responsible for determining the quality of work executed. Where work which has been included in a certificate subsequently proves to be defective, the value can be omitted from the next certificate. This may result in a negative value, but it is suggested that the effect of this will be that the certificate will show a payment due of 'zero', rather than a payment from the contractor to the employer (this is a difficult point of law, but it should be noted that the HGCRA 1996 (as amended) does not appear to contemplate 'negative' payments; see section 110A(4) of the Act). Note, however, that a certificate must be issued, even if the amount it shows is 'zero' (cl 4.5.6).

***Townsend* v *Stone Toms & Partners* (1984) 27 BLR 26 (CA)**

Mr Townsend engaged architects Stone Toms in connection with the renovation of a farmhouse in Somerset. John Laing Construction Ltd was employed to carry out the work on a JCT67 Fixed Fee Form of Prime Cost Contract. Following the end of the defects liability period, the architect issued an interim certificate that included the value of work which had already been included in the schedule of defects, and which the architect knew had not yet been put right. Mr Townsend brought proceedings against both Stone Toms and Laing. The Deputy Official Referee found that the architect was not negligent in issuing the interim certificate. Mr Townsend appealed and the Court of Appeal held that the architect had been negligent. Oliver LJ stated (at page 46):

> The whole purpose of the certification is to protect the client from paying to the builder more than the proper value of the work done, less proper retention, before it is due. If the architect deliberately over-certifies work which he knows has not been done properly, this seems to be a clear breach of his contractual duty, and whether certification is described as 'negligent' or 'deliberate' is immaterial.

Sutcliffe v *Chippendale & Edmondson* (1971) 18 BLR 149

(Note: this case is the first instance decision, which was appealed to the Court of Appeal *sub nom. Sutcliffe* v *Thackrah*, discussed at paragraph 7.1.)

Mr Sutcliffe engaged the architect Chippendale & Edmondson in relation to a project to build a new house. No terms of engagement were agreed, but the architect proceeded to design the house, invite tenders and arrange for the appointment of a contractor on JCT63. Work progressed slowly and towards the end of the work it became obvious that much of the work was defective. The architect had issued ten interim certificates before Mr Sutcliffe entirely lost confidence, dismissed the architect and threw the contractor off the site. He then had the work completed by another contractor and other consultants at a cost of around £7,000, in addition to which he was obliged, as a result of the original contractor having obtained judgment against him, to pay all ten certificates in full. As this contractor was then declared bankrupt, Mr Sutcliffe brought a claim against the architect. The architect contended, among other things, that its duty of supervision did not extend to informing the quantity surveyor of defective work that should be excluded from the valuation. HH Judge Stabb QC found for Mr Sutcliffe, stating 'I do not expect that the words "work properly executed" can include work not then properly executed but which it is expected, however confidently, the contractor will remedy in due course' (at page 166).

Dhamija and another v *Sunningdale Joineries Ltd and others* [2010] EWHC 2396 (TCC)

The claimants brought an action against the building contractor, the architect and the quantity surveyor (McBains) arising out of alleged defects in the design and construction of their home. There had been no written or oral contract with the quantity surveyor, but the claimants argued that there was an implied term that the quantity surveyor would only value work that had been properly executed by the contractor and was not obviously defective. The court held that a quantity surveyor's terms of engagement would include an implied term that the quantity surveyor act with the reasonable skill and care of a quantity surveyor of ordinary competence and experience when valuing the works properly executed for the purposes of interim certificates. However, the judge held that the quantity surveyor would not owe an implied duty to exclude the value of defective works from valuations, however obvious the defects. This was the exclusive responsibility of the architect appointed under the contract. Further, the quantity surveyor owed no implied duty to report the existence of defects to the architect.

Accepting defective work

7.8 As discussed in Chapter 5, it is possible to allow defective work to remain. Care should be taken when doing this. The contract administrator should obtain both parties' agreement, and ensure that its proposed adjustment to the contract price is agreed and confirmed in writing. The rates and prices for that work as set out in the contract can be used as a starting point for negotiation, but are not the only matters to be taken into consideration (see *Mul* v *Hutton Construction Limited*). The contractor may prefer to correct the work, especially if the proposed reduction in the contract price is significant, and in general it cannot be denied the opportunity to do so.

Mul v *Hutton Construction Limited* [2014] EWHC 1797 (TCC)

This case concerned what constitutes an 'appropriate deduction' when an employer decided to accept non-conforming work. The project concerned an extension and refurbishment work to a country house using the JCT IC05 form. A practical completion certificate was issued with a long list of defects attached, and during the rectification period the employer decided to have this work corrected by other contractors. The employer then started court proceedings against the contractor, to claim back the costs of this work.

A key issue was the interpretation of clause 2.30, which provides that the contract administrator can instruct the contractor not to rectify defects and 'If he does so otherwise instruct, an appropriate deduction shall be made from the Contract Sum in respect of the defects, shrinkages or other faults not made good'. In this case the contractor argued that an 'appropriate deduction' was limited to the relevant amount in the contract rates or priced schedule of works. The court disagreed. It decided that 'appropriate deduction' under clause 2.30 meant 'a deduction which is reasonable in all the circumstances', and could be calculated by any of the following: the contract rates or priced schedule of works; the cost to the contractor of remedying the defect (including the sums to be paid to third party sub-contractors engaged by the contractor); the reasonable cost to the employer of engaging another contractor to remedy the defect; or the particular factual circumstances and/or expert evidence relating to each defect and/or the proposed remedial works.

However, the judge also pointed out that the employer will still have to satisfy the usual principles that apply to a claim for damages, which include showing that it mitigated its loss. If the employer unreasonably refused to let the contractor rectify defects, then it is likely to find its damages limited to what it would have cost the contractor to put them right.

Unfixed materials and goods

7.9 Certificates should include for materials which have been delivered to the site but not yet incorporated in the works (cl 4.3.2). Although obliged under the contract to include such items, the contract administrator should be aware that this could result in some considerable risk to the employer. Once materials have been built in, under common law they would normally become the property of the owner of the land, irrespective of whether or not they have been paid for by the contractor. This would be the case even if there were a retention of title clause in the contract with the sub-contractor or supplier. A retention of title clause is one which stipulates that the goods sold do not become the property of the purchaser until they have been paid for, even if they are in the possession of the purchaser.

7.10 The employer could be at risk, however, where materials have not yet been built in, even where the materials have been certified and paid for. The contractor might not actually own the materials paid for because of a retention of title clause in the sale of materials contract. Under the Sale of Goods Act 1979, sections 16–19 (also referred to in the Consumer Rights Act 2015, section 4), property in goods normally passes when the purchaser has possession of them, but a retention of title clause will be effective between a supplier of goods and a contractor even where the contractor has been paid for the goods, provided they have not yet been built in. The employer may have some protection through section 25 of the Act, which in some circumstances allows the employer to treat the contractor as having authority to transfer the title in the goods (*Archivent* v *Strathclyde Regional Council*).

Archivent Sales & Developments Ltd v *Strathclyde Regional Council* (1984) 27 BLR 98 (Court of Session, Outer House)

Archivent agreed to sell a number of ventilators to a contractor who was building a primary school for Strathclyde Regional Council. The contract of sale included the term 'Until payment of the price in full is received by the company, the property in the goods supplied by the company shall not pass to the customer'. The ventilators were delivered and included in a certificate issued under the main contract (JCT63), which was paid. The contractor went into receivership before paying Archivent, which claimed against the Council for the return of the ventilators or a sum representing their value. The Council claimed that section 25(1) of the Sale of Goods Act 1979 operated to give it unimpeachable title. The judge found for the Council. Even though the clause in the sub-contract successfully retained the title for the sub-contractor, the employer was entitled to the benefit of section 25(1) of the Sale of Goods Act 1979. The contractor was in possession of the ventilators and had ostensible authority to pass the title on to the employer, who had purchased them in good faith.

7.11 Another risk relating to rightful ownership is where the contractor fails to pay a sub-contractor who has purchased and delivered materials for the works it will undertake, and the sub-contractor subsequently claims ownership of the unfixed materials. Here the risk may be higher, as a work and materials contract is not governed by the Sale of Goods Act 1979 (*Dawber Williamson Roofing Ltd* v *Humberside County Council*). Therefore there can be no assumption that property would pass on possession.

Dawber Williamson Roofing Ltd v *Humberside County Council* (1979) 14 BLR 70

The plaintiff entered into a sub-contract with Taylor and Coulbeck Ltd (T&C) to supply and fix roofing slates. The main contractor's contract with the defendant was on JCT63. By clause 1 of its sub-contract (which was on DOM/1) the plaintiff was deemed to have notice of all the provisions of the main contract, but it contained no other provisions as to when property was to pass. The plaintiff delivered 16 tons of roofing slates to the site, which were included in an interim certificate, which was paid by the defendant. T&C then went into liquidation without paying the sub-contractor, which brought a claim for the amount or, alternatively, the return of the slates. The judge allowed the claim, holding that clause 14 of JCT63 could only transfer property where the main contractor had a good title. (The difference between this and the Archivent case cited above is that in this case the sub-contract was a contract for work and materials, to which the Sale of Goods Act 1979 did not apply.)

7.12 Unlike IC11 or SBC16, MW16 contains no provisions to protect the employer from these risks. Nevertheless, under MW16 the contract administrator is obliged to include the unfixed materials when certifying interim payments. Clause 4.3.2, however, states that the obligation only extends to materials that are 'reasonably and properly' delivered and are adequately stored and protected. Contract administrators should pay careful attention to the exact wording of this qualification.

Retention

7.13 Interim payments are subject to a percentage deduction (often termed 'retention'), of which half is released after practical completion (cl 4.3). The percentage certified is to be

95 per cent (i.e. a 5 per cent reduction) unless another amount is inserted in the contract particulars. It would be open to the contract administrator to insert a lower figure, provided that this has been made clear to both the employer and the contractor at the time of tender. This figure might reflect the size and nature of the job, or the economic climate of the time. The 5 per cent retained on certain small projects is likely to be so small as to be almost worthless should the contractor become bankrupt or unexpectedly withdraw from site. If objections are raised at tender stage, it may be worth enquiring what the resulting difference is in the tender figure owing to the higher rate of retention.

7.14 The employer is not stated to be trustee for the beneficiaries of the retention, and there is no obligation for the employer to place the retained sum in a separate bank account. It is sometimes asserted that the contractor has a 'moral' right to this money, but it is difficult to see why this should be the case. It is simply a commercial arrangement to protect the employer, to which the contractor agrees in advance with full knowledge of the consequences. The contract administrator should resist any suggestion that the money be treated in any way other than that set out in the contract itself.

Payment procedure

7.15 Once an interim payment is certified, the final date for payment is 14 days from the due date (cl 4.3) (Figure 7.1). It should be noted that this has changed from earlier versions of the form, where the 14 days ran from the date of the certificate; so, unless the certificate is issued on the due date, the time period before payment is required will be reduced. As this timing is rather tight, it would be sensible to issue the certificate promptly and to warn the employer in advance of approximately what payment may be required. It is also helpful to include with the certificate a budget update showing projected total contract sum based on any information known at that time regarding variations and provisional sums. This helps the employer to make arrangements for payment in good time.

7.16 If the employer intends to withhold any amount from the sum certified, the employer (or the contract administrator, on the employer's behalf) must give the contractor written notice of this no later than five days before the final date for payment (cl 4.5.4.1). The notice (termed a 'pay less notice') should set out the sum that the employer considers is due to the contractor at the date the notice is given and 'the basis on which that sum has been calculated' (cl 4.5.4). If a notice is issued, the employer must pay the contractor at least the sum set out in that notice by the final date for payment.

Payment when no certificate is issued

7.17 The contract has a remedy should the date for issue pass without a certificate having been provided. If the contractor has already issued an application for payment, then this becomes a 'payment notice' (cl 4.4.2.1). Otherwise, the contractor can, at any time after the issue date has passed, send a payment notice to the contract administrator (cl 4.4.2.2). The notice should state 'the sum that the Contractor considers to have become due to him under clauses 4.3 or 4.8 at the relevant due date and the basis on which that sum has been calculated'. The employer must then pay the contractor the sum shown as due on the payment notice, subject to any pay less notice (cl 4.5.2),

Figure 7.1 Interim payment procedure

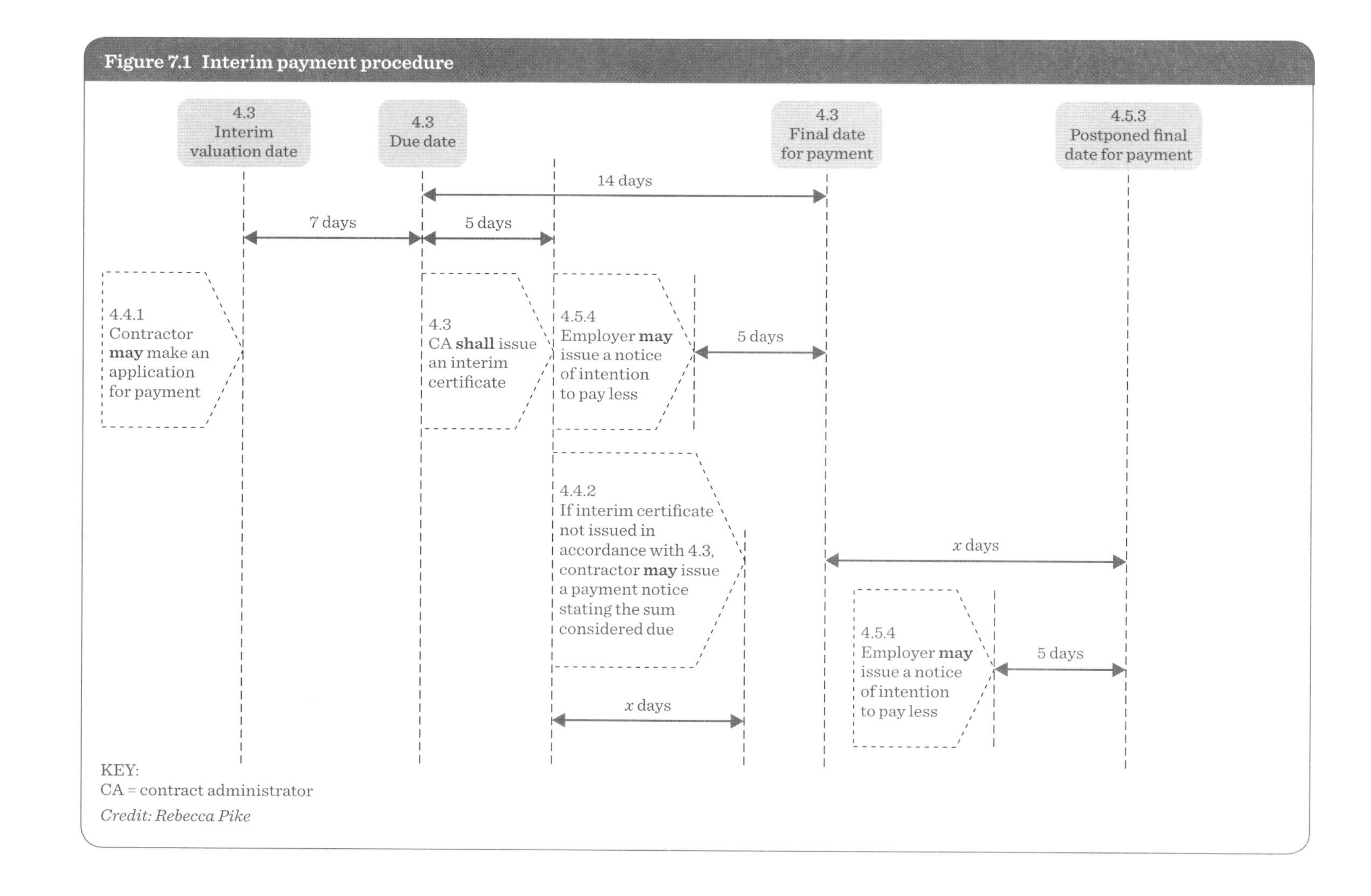

7.18 Where a payment notice is given, clause 4.5.3 states that 'the final date for payment of the sum specified in it shall for all purposes be regarded as postponed by the same number of days as the number of days after expiry of the 5 day period referred to in clause 4.4.2.2 that the Contractor's payment notice is given'. This is best understood by way of example; if the payment notice was issued six days after the final date for issue of the certificate (i.e. 11 days in total after the due date), the final date for payment would be 20 days after the due date (i.e. 14 plus six days). If the employer disagrees with the amount shown on the payment notice then it must issue a notice explaining this intention, as described above (which in this example would be within four days of the payment notice).

Deductions

7.19 The contract expressly allows the employer to deduct liquidated damages from certified amounts (cl 2.8.1, or 2.9.1 in MWD16), provided the correct notices are issued. Other deductions are to be made from the contract sum and taken into account when calculating the amount shown on certificates (see paragraph 7.6). Nevertheless, it is suggested that, should the contract administrator have inadvertently omitted to make a deduction allowed under the contract, the employer would be entitled to do so by means of a notice.

Employer's obligation to pay

7.20 It is now generally agreed that in cases where the employer has a right to make a deduction, this can only be exercised through the use of the 'notice' procedure as discussed above. The employer would therefore be unable to withhold amounts to cover any defective work included in a certificate unless the deduction is covered by a notice (*Rupert Morgan Building Services (LLC) Ltd* v *David Jervis and Harriet Jervis*). The rights of the employer when defects appear after the expiry of the time limits for notices but before the final date for payment are unclear, but it is arguable that in such situations the employer would retain a right to abatement of the amount due.

Rupert Morgan Building Services (LLC) Ltd v *David Jervis and Harriet Jervis* [2003] EWCA Cir 1583 (CA)

A couple engaged a builder to carry out work on their cottage, by means of a contract on the standard form published by the Architecture and Surveying Institute ('ASI'). The 7th interim certificate was for a sum of around £44,000 plus VAT. The employers accepted that part of that amount was payable but disputed the balance, amounting to some £27,000. The builder sought judgment for the balance. The employers did not give 'a notice of intention to withhold payment' before 'the prescribed period before the final date for payment'. The builder argued that it followed, by virtue of section 111(1) that the employers 'may not withhold payment'. The employers maintained that it was open to them by way of defence to prove that the items of work which go to make up the unpaid balance were not done at all, or were duplications of items already paid or were charged as extras when they were within the original contract, or represent 'snagging' for works already done and paid for. The Court of Appeal determined that in the absence of an effective withholding notice the employer has no right to set off against a contract administrator's certificate.

Contractor's position if the certificate is not paid

7.21 MW16 includes several provisions which protect the contractor if the employer fails to pay the contractor amounts due. Clause 4.6.1 makes provision for simple interest on late payment of interim and final certificates. This is set at 5 per cent above the official bank rate of the Bank of England (cl 1.1), and the interest accrues from the final date for payment until the amount is paid. If the employer makes a valid deduction following a notice it is suggested that interest would not be due on this amount. The clause does not refer to the amount stated on the certificate but to 'a sum ... due to the other Party', which would take into account valid deductions. The contractor may recover the interest as a debt from the employer. Clause 4.6.2 makes it clear that acceptance of payment of interest does not limit the contractor's rights to take any other action under the contract to recover the principal amount.

7.22 The contractor is also given a 'right of suspension' under clause 4.7. This right is required by the HGCRA 1996 (as amended). If the employer fails to pay the contractor by the final date for payment, the contractor has a right to suspend performance of all its obligations under the contract, which would not only include the carrying out of the work, but could also, for example, extend to any insurance obligations. This right results from a failure by the employer to pay 'the sum payable to the Contractor in accordance with clause 4.5', which suggests that the contractor may not suspend work if a notice to withhold payment has been given by the employer. The contractor must have given the employer written notice of its intention to suspend work and stated the grounds for the suspension, and the default must have continued for a further seven days (cl 4.7.1).

7.23 The contractor must resume work when the payment is made. Under these circumstances the suspension would not give the employer the right to terminate the contractor's employment. Any delay caused by the suspension could be a matter beyond the control of the contractor in relation to an extension of time. In addition, the contractor is entitled to any reasonable 'costs and expenses' incurred due to the suspension (cl 4.7.2), and should send details of these to the contract administrator for inclusion under the next certificate (cl 4.7.3).

7.24 The contractor also has the right to terminate the contract if the employer does not pay amounts due in accordance with clause 4.5 (cl 6.8.1.1). The contractor must give notice of this intention, which specifies the default as required by the contract.

Contractor's position if it disagrees with an amount certified

7.25 The issue of a certificate is not a condition precedent to the right of the contractor to be paid, as in its absence the contractor can issue a clause 4.4.2 payment notice. However, where a certificate has been issued, the contractor is only entitled to the amount shown, even if it is undervalued (*Lubenham* v *South Pembrokeshire District Council*). This case states that the contractor is only entitled to the sum stated in the certificate, even if the certificate contains an error, for example because it includes a wrongful deduction. The contractor's remedy is to request that the error is corrected in the next certificate, or to bring proceedings to have the certificate adjusted. (There are exceptions to this rule where, for example, the employer has interfered with the issue of the certificate, in which case the contractor may be entitled to summary judgment for the correct amount.)

Lubenham Fidelities and Investments Co. Ltd v *South Pembrokeshire District Council* (1986) 33 BLR 39 (CA)

Lubenham Fidelities was a bondsman which elected to complete two building contracts, both based on JCT63. The architect, Wigley Fox Partnership, issued several interim certificates which stated the total value of work carried out, but also made deductions for liquidated damages and defective work from the face of the certificate. Lubenham protested that the certificates had not been correctly calculated, withdrew its contractors from the site and issued notices to determine the contract. Shortly afterwards, the Council gave notice of determination of the contract. Lubenham brought a claim against the Council on the grounds that its notices were valid and effective, and against Wigley Fox on the basis that the architect's negligence had caused it losses. It was held that the Council was not obliged to pay more than the amount on the certificate, and that whatever the cause of the undervaluation the correct procedure was not to withdraw labour, but to request that the error was corrected in the next certificate, or to pursue the matter in arbitration. Lubenham's claim against Wigley Fox failed because it was the suspension of the works rather than the certificates that had caused the losses, and because the architect had not acted with the intention of interfering with the performance of the contract.

Certificates after practical completion

7.26 Under MW16 certificates continue to be issued at the same intervals after practical completion up until the final payment (cl 4.3; note there is no longer a separate clause covering a 'penultimate certificate'). Following practical completion, the amount should be for 97.5 per cent (unless a different percentage was inserted in the contract particulars) of the total amount due to the contractor under the contract, in so far as it can be ascertained at the valuation date. The effect would therefore normally be to release to the contractor half of the retention that has been deducted. The precise final amount due to the contractor might not be known until final payment stage, but a reasonably accurate calculation should be made at each valuation based on all available information. As above, a certificate should still be issued, even if the amount payable is 'zero'. The employer is obliged to pay within 14 days of the due date, and the provisions regarding notices apply also to these certificates. As with all interim certificates, if a certificate is issued late, this will effectively reduce the time interval within which the payment must be made.

Final certificate

7.27 To summarise, by final certificate stage the following certificates should have been issued:

- interim payment certificates at monthly intervals (cl 4.3);
- practical completion certificate (cl 2.9, or 2.10 in MWD16);
- interim payment certificates following practical completion, including release of half of the retention (cl 4.3);
- certificate of making good (cl 2.11, or 2.12 in MWD16).

7.28 The date for issuing the final certificate will be determined by whether the contract administrator has certified that the contractor has discharged its obligations concerning defects at the end of the rectification period, and by whether the contractor has supplied

Figure 7.2 Final payment procedure

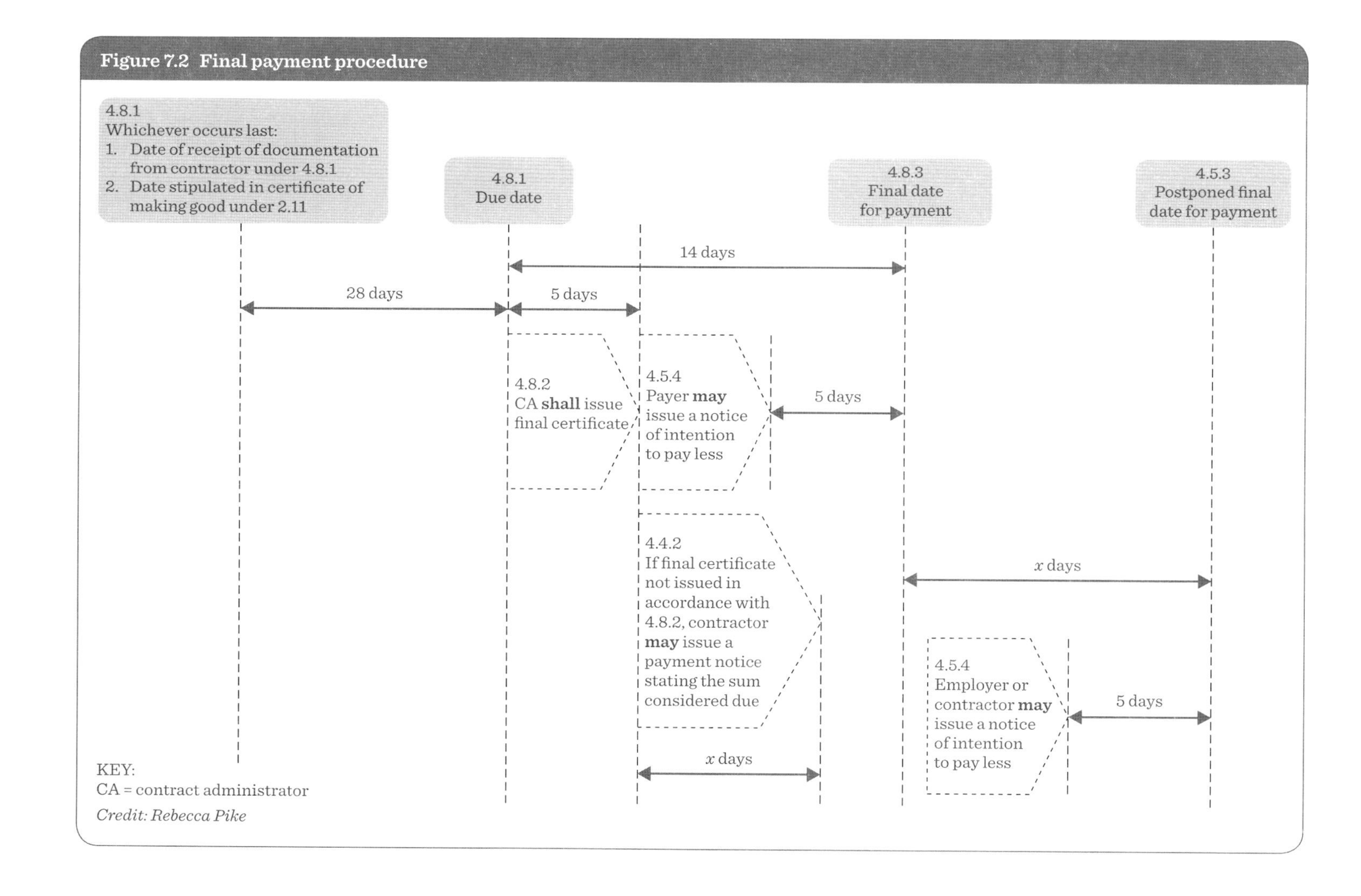

KEY:
CA = contract administrator

Credit: Rebecca Pike

sufficient documentation for the preparation of the final account (Figure 7.2). The contractor must send the information 'reasonably required' within three months of the date of practical completion certified by the contract administrator or other period stated in the contract particulars (cl 4.8.1). Where the contract administrator finds that the information is inadequate, then any outstanding information should be asked for immediately. The due date for final payment is 28 days after either the date of receipt of the documentation 'reasonably required' or, if later, the date specified in the certificate under clause 2.11 (2.12 in MWD16) (making good). The contract administrator must then issue the final certificate within five days of the due date. Although it might be considered good practice to allow the contractor time to consider the draft final account, there is no such requirement and the time-scale could be tight in some instances.

7.29 The final certificate must state the basis of calculation and the amount remaining due to the contractor. Clause 4.8.2.1 sets out the adjustments that should be taken account of in assessing the final contract sum, which comprises: the value of 'work properly executed' (assessed as per interim certificates), any fluctuations adjustments, and any relevant deductions as allowed under clause 2.10 (defects not corrected in rectification period) or clause 3.5 (non-compliance with instructions). The certificate should also state the total amount shown as due in previous interim certificates, and any payment made by the employer as a result of a contractor's payment notice. The balance due could, in unusual circumstances, be for a negative amount – in other words, it could certify that payment is due from the contractor to the employer. The certificate is subject to the same rights to make deductions as are discussed in paragraph 7.19. The time for final payment is 14 days from the due date. As with interim certificates, if the final certificate is issued late, this will effectively reduce the time interval within which the final payment must be made.

Conclusive effect of final certificate

7.30 The final certificate is not stated to be conclusive evidence that any obligations under the contract have been discharged. It is important to point this out, as this is not the case in other JCT forms, such as SBC16. The conclusive effect of the final certificate was the subject of much heated debate following the decisions in *Colbart Ltd* v *H Kumar* and *Crown Estate Commissioners* v *John Mowlem*. These cases did not concern the Minor Works Building Contract, where the wording of the relevant clauses is quite different, and the approach taken in *Crown Estate Commissioners* v *John Mowlem* has not been applied to that form. Earlier cases which did consider the Minor Works Agreement, such as *Crestar* v *John and Joy Carr*, reached a quite different conclusion as to the effect of the final certificate.

Crestar Ltd v *Michael John Carr and Joy Carr* (1987) 37 BLR 113 (CA)

John and Joy Carr employed Crestar on the Minor Works Form (pre-1980 version) to carry out work on their house. The contract price was £70,634. Crestar claimed it had carried out additional work to the value of around £46,000. The work was finished about June 1985, and on 1 October 1985 the architect issued a final certificate in the sum of £39,575, which valued the additional work at £49,690. Prior to the final certificate, the owners had paid all amounts certified, but they did not pay this amount. On 22 October Crestar issued a writ for the amount certified and the Carrs applied for a stay, on the grounds that the contract contained an arbitration clause.

The Carrs' application was granted by the district registrar and Crestar appealed. The court dismissed the appeal and Crestar then appealed again, claiming that the effect of the final certificate was that the employer could not bring a claim in arbitration after 14 days from the date of its issue as it then constituted a debt due, and the certificate became final and conclusive as to the quality of the works. Fox LJ, dismissing the appeal, pointed out that the conditions assumed that by the time of the final certificate the employer would normally have been presented with a certificate and have paid 95 per cent of the amount due following practical completion. As no penultimate certificate had been issued, the certificate issued in October was not a valid final certificate under the terms of the contract.

However, even if it had been valid there was nothing in the contract that made it conclusive of any matters. He stated (at page 124):

> There is nothing in the arbitration provisions to prevent it [the employer raising a defence]. There is nothing in any other Conditions of the contract to prevent it save, it is said, the word 'debt'. I do not think that is sufficient: it places undue weight upon the term ... I am not prepared to infer that the parties intended to prevent the owners referring matters to arbitration after the end of the 14-day period. Important defects in work or materials which were not apparent upon reasonable inspection prior to the expiry of the 14-day period may become apparent later.

7.31 It is nevertheless worth pointing out that although neither the final certificate nor any other certificate is stated to be conclusive, there may remain a very limited area where it would be difficult to raise a challenge. This could occur where the contract administrator has not specified any objective standards against which a decision could be reviewed, and no such standards could be implied. For example, if the contract administrator has merely indicated that the colour of the paint should be 'apple green' and 'to the satisfaction of the architect', if the contract administrator includes for painting work in a certificate, it may be very difficult for the employer to later claim that the colour of the paint is unsatisfactory. Although in theory the certificate could be challenged, there would be no means by which it could be proved the contractor was in breach. The message is clear – always specify objective and measurable standards.

8 Indemnity and insurance

8.1 Even small-scale operations carry risks, particularly those relating to work on existing buildings. It is often the case with such work, where the programme may be tight and arrangements anything but straightforward, that matters of insurance are not dealt with as promptly or as thoroughly as they might be. It is important that liability for losses resulting from personal injury or damage to property or to the works is agreed beforehand and clearly allocated to one party or the other, and that the liability is backed up by appropriate insurance cover. In seeking to protect the employer's interests the contract administrator should have sufficient knowledge of the relevant contract clauses to be able to explain their purpose and what they mean.

8.2 Most building contracts provide for the contractor to indemnify the employer in respect of certain losses, for example for injury to persons, or damage to neighbouring property which has been caused by the contractor's negligence, and in MW16 this is done under clauses 5.1 and 5.2. This indemnity protects the employer in that if an injured party brings an action against the employer, rather than against the contractor, the latter has agreed to carry the consequences of the claim. If a third party sues the employer, then the employer can join the contractor as co-defendant or bring separate proceedings. Indemnities given to the employer by the contractor will obviously be quite worthless unless there are adequate resources to meet claims. The contract therefore requires insurance cover to back up the indemnities.

8.3 In addition to the requirement for insurance against claims arising in respect of persons and property, the contract also contains alternative provisions for insurance of the works under clauses 5.4A, 5.4B and 5.4C. Clause 5.4A is for use with new buildings, and is taken out by the contractor, 5.4B is for use with existing buildings where the policy will cover the new works and existing structure, and is taken out by the employer, and 5.4C is used where neither of these options is appropriate and the parties agree their own insurance arrangements (these are to be set out in the contract particulars). It should be noted that MW16 makes no provision for terrorism cover or for compliance with the Joint Fire Code.

8.4 Neither does MW16 include provisions for insurance against damage caused to property which is not the result of the negligence of the contractor. For example, subsidence or vibration resulting from the carrying out of the works might cause such damage, even though the contractor has taken reasonable care. This is a risk which may be quite high with certain projects, such as those on tight urban sites or in close proximity to old buildings, and in such cases it may be advisable to take out a special policy for the benefit of the employer. If this type of insurance is desired then it may be possible to import an adapted wording of the appropriate clause in SBC16 or IC11, requiring the contractor to take out such insurance, but specialist advice should be sought. This insurance is usually expensive, and subject to a great many exceptions. If it is required then the policy needs to be effective at the start of the site operations when demolition, excavation, etc. are carried out.

8.5 Both parties must be able to provide reasonable evidence within seven days of a request that the insurances referred to under clauses 5.3 and 5.4A, 5.4B or 5.4C have been taken out (cl 5.5). In cases where an existing policy is to be extended to cover the other party, that party will need to check carefully that the policy has been appropriately endorsed, recognises them as a joint insured and provides exactly the right cover. There is no provision for what will occur if either party fails to provide evidence or defaults on their insurance obligations; in such cases the other party may need to take out the insurance and bring a claim to recover its losses.

Injury to persons

8.6 Clause 5.1 covers injury to persons which arises from the carrying out of the works. The contractor is liable and indemnifies the employer, but the liability and duty to indemnify are subject to the exception that the contractor is not liable where injury or death is caused by an act of the employer or an employer's person or by any statutory undertaker (cl 5.1).

8.7 The contractor's liability in respect of personal injury or death of employees is met by an employer's liability policy. This has been compulsory since the Employers' Liability (Compulsory Insurance) Act 1969. The legal minimum level of cover for most firms is £5 million, but many insurers will provide a £10 million policy as standard.

8.8 The contractor's liability in respect of third parties (death or personal injury and loss or damage to property including consequential loss) is met by its public liability policy. Insurers typically advocate insuring for a minimum of £2 million for any one occurrence, and insurance companies typically offer £5 million as the standard level of cover. Liability at common law for claims by third parties is, however, unlimited. There is no opportunity to enter a minimum requirement in the contract but the policy must comply with 'all relevant legislation' (cl 5.3.1), which would also cover, for example, insurance requirements under the Road Traffic Act 1988.

Damage to property

8.9 Clause 5.2 covers damage to property, real or personal, which arises from the carrying out of the works. The works, including unfixed goods and materials, are expressly excluded from the definition of property in this context. The contractor is required to match the indemnity required under clause 5.2 with appropriate insurance (cl 5.3).

8.10 The minimum figure for which the contractor is required to take out insurance cover should be entered in the contract particulars (clause 5.3.2), and is unlikely to be less than £2 million. This would be a contractual minimum and in no way limits the contractor's liability to the sum entered.

8.11 The contractor is only liable to the extent that the damage is caused by negligence or breach of statutory duty or other default of 'the Contractor or any Contractor's Person' (cl 5.2). The contractor is therefore liable only for losses caused by its own negligence.

8.12 Clause 5.2 also excludes, where clause 5.4B applies, liability for 'any loss or damage to those existing structures or to any of their contents that are required to be insured under clause 5.4B.1 that is caused by any of the risks or perils required or agreed to be insured against under clause 5.4B'. This means that, where clause 5.4B is applicable, the contractor is not liable for losses insured under clause 5.4B.1 and caused by the listed perils, even where the damage is caused by the contractor's own negligence. This point is now expressly stated in clause 5.4B.2. The exclusion was inserted in an earlier edition of the form to clarify matters following a series of cases on older versions of the contract that reached the opposite conclusion (*National Trust* v *Haden Young, Barking & Dagenham* v *Stamford Asphalt Co.*). Domestic sub-contractors, however, may be liable for losses caused by their negligence (*BT* v *James Thompson & Sons*). It should be noted, also, that the contractor might remain liable for some consequential losses (*Kruger Tissue* v *Frank Galliers*), and for damage caused to the works and insured under clause 5.4B.2.

***The National Trust for Places of Historic Interest and Natural Beauty* v *Haden Young Ltd* (1994) 72 BLR 1 (CA)**

The National Trust employed a contractor to carry out repair works to Uppark House, South Harting, West Sussex. The main contract was on terms substantially similar to MW80. Haden Young was sub-contractor for the renewal of lead work on the roof. During the course of the works a fire broke out, causing extensive damage, which Haden Young admitted was caused by the negligence of its workforce, and the National Trust brought a claim for damages. Otton J found the sub-contractor liable at first instance, and that the employer's liability to insure under clause 5.4B only extended to matters not caused by negligence. Clauses 5.2 and 5.4B formed a coherent and mutually supportive structure. Haden Young appealed, but the appeal was dismissed. Although the Court of Appeal agreed that the sub-contractor was liable, it disagreed with the reasoning of the lower court, stating that there was no reason why there should not be an overlap, in other words why the employer should not be required to insure for matters for which the contractor was liable under clause 6.2. However, the damages recoverable from the contractor under clause 6.2 would be reduced by the amount recoverable by the employer under the clause 6.3B insurance.

***London Borough of Barking & Dagenham* v *Stamford Asphalt Co. Ltd* (1997) 82 BLR 25 (CA)**

Barking & Dagenham employed a contractor to carry out repair works to a school. The main contract was on MW80, 1988 revision. Stamford was sub-contractor for the renewal of lead work on the roof. During the course of the works a fire broke out, causing extensive damage, which Stamford admitted was caused by the negligence of its workforce, and the Borough brought a claim for damages. The Court of Appeal found the contractor liable for the damage caused, preferring the reasoning of Otton J in ***National Trust* v *Haden Young*** to that of the Court of Appeal in that case. It should be noted that the wording of clause 6.2 (now cl 5.2) has been adjusted to make it clear that the contractor is not liable for damage to property insured under clause 6.3B (now cl 5.4B).

***British Telecommunications plc* v *James Thompson & Sons (Engineers) Ltd* (1999) 1 BLR 35 (HL)**

James Thompson was a sub-contractor on a refurbishment project for BT. A fire broke out in the roof area while the sub-contractor was carrying out its work. The court found that the relevant clauses had the same effect as the equivalent clauses considered in *Scottish Special Housing Association* v *Wimpey*. However, it decided that domestic sub-contractors remained under a duty of care to prevent such losses, and James Thompson was therefore liable to BT under the tort of negligence. They considered that the wording of clause 22.3, which required the joint names policies to waive the rights of subrogation against nominated but not domestic sub-contractors, should be taken into account in considering whether a duty of care existed. The fact that BT was indemnified by the clause 22C insurers, even if the fire was caused by the sub-contractor, was insufficient to prevent the imposition of the duty.

Kruger Tissue (Industrial) Ltd v *Frank Galliers Ltd* (1998) 57 Con LR 1

Damage was caused to existing buildings and the works by fire, assumed for the purposes of the case to be the result of the negligence of the contractor or sub-contractor. The construction work being carried out was on a JCT80 form. The employer brought a claim for loss of profits, increased cost of working and consultants' fees, all of which were consequential losses. Judge John Hicks decided that the employer's duty to insure for 'the full cost of reinstatement, repair or replacement of the existing structure and the Works under clause 22C (and therefore contractor's exemption from liability under clause 20.2), did not include such consequential losses'. A claim could therefore be brought against the contractor for these losses.

Insurance of the works

8.13 There are three alternative clauses for insuring the works (cl 5.4A, 5.4B and 5.4C) and the reference to the clauses which are not applicable should be deleted in the contract particulars.

8.14 The policies under clause 5.4A or 5.4B are to be in joint names ('Joint Names Policy' is defined under clause 1.1), and the insurer waives any rights to recover any of the monies from either of the named parties or from any person recognised as insured under the policy. The cover must run until practical completion of the works, or termination if this should occur earlier.

8.15 Clause 5.4A is intended for insuring new building work, with no existing structures involved. It requires the contractor to take out 'All Risks' insurance against loss or damage to the works and unfixed goods and materials (the term 'All Risks Insurance' is defined under clause 1.1; this insurance is often referred to as 'CAR', contractor's all risks). The insurance should be for the full reinstatement value of the works (which is likely to be more than the original tender figure). It must include for professional fees likely to arise should damage occur, at the percentage entered in the contract particulars.

8.16 The contractor is responsible for keeping the works fully covered, and in the event of underinsurance the contractor will be liable for any shortfall in recovery from the insurers. Care needs to be taken over identifying the extent of 'the Works' exactly.

8.17 Clause 5.4B is applicable where work is being carried out to existing buildings. The existing structure together with its contents owned by the employer 'or for which he is responsible' are to be insured in joint names against damage due to the 'Specified Perils' (effectively the employer's existing policy will be extended to cover the contractor against the specific losses set out in 5.4B.1). The employer additionally takes out a joint names 'All Risks' policy to cover the works (5.4B.2). As noted above, the insurance under clause 5.4B.1 (but not 5.4B.2) will protect the contractor even if the damage results from its own negligence. Sometimes the employer has difficulty in obtaining this insurance: either the building may be historic or particularly sensitive and therefore extremely difficult to insure, or, in the case of small domestic projects, the insurers may refuse to extend cover when they are informed about impending work. Insurance under clause 5.4B will be particularly difficult if the employer is not the freeholder.

8.18 In the event that the employer cannot obtain insurance under clause 5.4B, the parties will need to make special arrangements under clause 5.4C. The JCT in its guidance notes suggests three options. Where the employer is the freeholder, the employer insures the works under joint names, and continues with its own existing structures policy. The contractor insures itself against damage to existing structures. Alternatively, the contractor insures the works under clause 5.4A, together with the existing structure (by extending its works insurance policy). Finally, if the employer is not the freeholder, then the JCT recommends that the parties take specialist advice, and liaise with the freeholder and its insurers.

8.19 Whatever arrangements are made under clause 5.4C, the employer will need to consider what these might be before tenders are sought, and there are likely to be negotiations before the matter can be finalised. All relevant insurers should of course be consulted, and (particularly for inexperienced employers) specialist advice may well be required. The contract administrator should raise insurance matters with the employer at an early point in the project, and draw their attention to the helpful JCT guidance.

Action following damage to the works

8.20 The contractor is required to inform the contract administrator and the employer as soon as any damage occurs (cl 5.6.1). The insurers should be immediately informed. The contractor is required to authorise the insurers to pay all monies under the works insurance policy direct to the employer (cl 5.6.3).

8.21 Where the damage to the works and site materials is covered by the works insurance policy, and after any inspection required has been made by the insurers, the contractor is then obliged to make good the damage and continue with the works (cl 5.6.4).

8.22 Where the contractor has effected the works insurance policy under clause 5.4A or 5.4C, the insurance monies paid to the employer, minus the part assigned to cover professional fees, should be included in separate reinstatement certificates as the work is carried out, issued at the same time as the interim certificates (cl 5.6.5). If the amount paid by the insurers is less than it costs the contractor to rebuild the works, the contractor is not entitled to any additional payment (cl 5.6.5.3). The risk of any underinsurance therefore lies with the contractor.

8.23 Where clause 5.4B applies, or the employer has effected the works insurance policy under clause 5.4C, or where loss is caused by an excepted risk (i.e. it is not insured), the contract administrator must issue instructions regarding the rebuilding work which is treated as if it were a variation under clause 3.6.1 (cl 5.6.6). It will therefore be paid for under the normal interim certificates. The contractor is less at risk as the employer will have to bear any shortfall in the monies paid out. In addition, as the work is treated as a variation, the contractor may be entitled to loss and/or expense.

8.24 Under clause 2.7 (2.8 in MWD16) the contractor might be entitled to an extension of time for delay caused by the events, except it is suggested that this would not extend to cases where the damage was caused by the contractor's negligence.

8.25 The employer is under no obligation to continue with the project where there is material (i.e. extensive) damage to the existing structure. Either party may in these circumstances terminate the contractor's employment by means of a 28-day notice (cl 5.7). If the other party feels that the project should continue, it must invoke the dispute resolution procedures, otherwise the clause 6.11 termination consequences will apply.

8.26 MW16 also contains provisions enabling either party to terminate the employment of the contractor in the event that work is suspended for a period of one month or more as a result of loss or damage to the works caused by any risk covered by the works insurance policy or by an excepted risk (cl 6.10.1.3). It is suggested that this right would normally only be exercised where the damage is so extensive as to make it impracticable or impossible to continue with the works; if the damage is limited it will normally be in both parties' interests to continue under the existing contract. It should be noted that the contractor is not entitled to terminate the contract if the specified peril was caused by its own negligence (cl 6.10.2).

The contract administrator's role in insurance

8.27 The insurance provisions in MW16 might appear to indicate no role for the contract administrator, but it would normally be implied that the contract administrator has a duty to explain the provisions in the contract to the employer. The contract administrator will keep a watch on the actions of the parties, although insurance matters may appear to be conducted almost directly between them. The contract administrator is primarily a channel of communication, and although responsibility for policies rests primarily with the contractor or the employer and their insurance advisers, the contract administrator should check the wording of policies and any endorsements to see that there are no obvious inconsistencies with the contract documents.

8.28 The contract administrator should therefore have a reasonable knowledge of the indemnity and insurance provisions of MW16, although would not be expected to be an expert. It may be that difficulties arise in obtaining cover for certain types of building, or that the cost of insurance becomes uneconomic and the employer wishes to adopt a policy of no insurance. It could be that insurers are reluctant to give joint names cover on existing domestic structures, or may well seek to impose special conditions. In particular, the relationship between the employer's existing insurance arrangements and those required under the contract can be very tricky, and on matters such as these the employer must be advised to consult its own insurance experts.

8.29 If accidents do occur on site and injury or damage results, the contract administrator must be alert to the fact that such matters must be reported quickly to the insurers. If damage to the works requires inspection by the insurers, then the contract administrator might be expected to be in attendance or to supply information. If loss or damage results which is the direct responsibility of the employer under clause 5.4B then the contract administrator is required to issue instructions for the reinstatement and to value the work as may be necessary.

9 Termination

9.1 At the outset of a contract the parties almost invariably approach it in good faith and with the best of intentions but, in spite of this, breaches of contract sometimes occur. Some of these are minor technicalities, and some can be dealt with by the machinery of the contract. Others are more serious and can only be dealt with by more drastic measures.

Repudiation or termination

9.2 Most building contracts include provisions to deal with foreseeable situations which might otherwise be breaches. For example, where the contractor is unable to complete by the completion date due to circumstances entirely beyond its control, then an extension of time can be awarded. Where the employer wishes to vary the work after the contract has been let, then this is possible under clause 3.6. However, if such machinery is not included in a contract, or is not operated as it should be, then the injured party may be able to claim damages for breach of contract. Such claims would have to go to adjudication, arbitration or litigation.

9.3 It sometimes happens that the behaviour of one party makes it difficult or impossible for the other to carry out its contractual obligations. The injured party might then allege prevention of performance and sue either for damages or a *quantum meruit*.

9.4 Alternatively, where it is impossible to expect any further performance, the injured party might allege that the contract has been repudiated. Repudiation occurs when one party makes it clear that it no longer intends to be bound by the provisions of the contract. This intention might be expressly stated, or implied by the party's behaviour.

9.5 Most JCT contracts include termination clauses, which provide for the effective termination of the employment of the contractor in circumstances which may amount to, or which may fall short of, repudiation. (If there is repudiation, invoking a termination clause is unnecessary, because the injured party can accept the repudiation and bring the contract to an end.) It should be noted that the termination is of the contractor's employment, and is not termination of the contract. In effect, termination removes the need for further performance, but leaves the parties still bound by other provisions. Termination provisions, such as those set out in MW16 Section 6, are useful in setting out the exact circumstances, procedures and consequences of the termination of employment. However, these procedures must be followed with great caution because if they are not administered strictly in accordance with the terms of the contract, this in itself could amount to repudiation. This, in turn, might give the other party the right to treat the contract as at an end and claim damages.

9.6 Under MW16 the employer has the right to terminate the contractor's employment in the event of specified defaults by the contractor, such as suspending the carrying out of the works (cl 6.4), or in the event of the insolvency of the contractor (cl 6.5), or in the event of corruption, for example that the contractor has committed an offence under the Bribery Act 2010 (cl 6.6). Termination can be initiated by the contractor in the event of specified defaults by the employer, such as failure to pay the amount due on a certificate or causing the work to be suspended for a period of more than one month (cl 6.8). Termination might also follow the insolvency of the employer. Termination may be initiated by either party in the event of neutral causes that might cause the work to be suspended for a period of one month or more (cl 6.10).

Termination by the employer

9.7 The contract provides for termination of the employment of the contractor under stated circumstances. MW16 expressly states that the right to terminate the contractor's employment is 'without prejudice to any other rights and remedies' (cl 6.3.1). This termination can be initiated by the employer in the event of specified defaults by the contractor occurring prior to practical completion (cl 6.4.2), which comprise suspending the carrying out of the works or, in the case of MWD16, the design of the CDP (cl 6.4.1.1), failing to proceed regularly or diligently with either of these (cl 6.4.1.2) and breach of the CDM Regulations (cl 6.4.1.3). The employer may also terminate in the event of insolvency, or corruption of the contractor (cl 6.5.1 and 6.6). Where the employer is a local or public authority, circumstances as set out in section 73(1)(b) of the Public Contracts Regulations 2015 (conviction of various offences) will also give rise to the right to terminate (cl 6.6). Upon termination the contractor must immediately leave the site; the employer need not make any further payment until the works are complete (cl 6.7.2), and may recover any losses resulting from the termination from the contractor (cl 6.7.3.1).

9.8 The procedures as set out in the contract must be followed exactly, especially those concerning the issue of notices under clause 6.4. If default occurs, the contract administrator should issue a warning notice, specifying the default and requiring it to be ended. If the default is not ended within seven days from receipt of the notice then the employer may terminate the employment of the contractor by the issue of a further notice within ten days from the expiry of that seven-day period, which takes effect from the date of receipt (cl 6.4.2). Unlike SBC16 and IC11, MW16 has no provision for 'repeat' defaults: if this occurs it is likely that a further warning notice will be needed. In the case of insolvency or corruption, only one notice is required to be issued by the employer (cl 6.5.1 and 6.6).

9.9 It should be noted that to be valid all notices must be in writing and given by actual delivery, or by special or recorded delivery (cl 6.2.3). It appears that this may include fax or email, provided that it can be proved the notice was received (*Construction Partnership* v *Leek Developments*). Notices sent by post are deemed to have been received 48 hours after posting (excluding weekends and public holidays), unless there is proof to the contrary. Regardless of the delivery method selected, as time limits are of vital importance it might be wise to have receipt of delivery confirmed.

Construction Partnership UK Ltd **v** ***Leek Developments Ltd*** **[2006] CILL 2357 (TCC)**

On an IFC98 contract, a notice of determination was delivered by fax, but not by hand or by special delivery or recorded delivery. (A letter had been sent by normal post but it was unclear whether or not it had been received.) Clause 7.1 required actual delivery of notices of default and determination, and the contractor disputed whether the faxed notice was valid. The court therefore had to decide what 'actual delivery' meant. It decided that it meant what it says: 'Delivery simply means transmission by an appropriate means so that it is received'. In this case, it was agreed that the fax had been received, therefore the notice complied with the clause. The CILL editors state that 'on a practical level, this judgement is quite important' because it had previously been assumed that 'actual delivery' meant physical delivery by hand. In their view, email could be considered an appropriate method of delivery, although that was not decided in the case.

9.10 The grounds for termination by the employer under clause 6.4.1 include failing to proceed diligently, wholly or substantially suspending the carrying out of work and failing to comply with obligations under the CDM Regulations. The grounds must be clearly established and expressed as the contract clearly states that termination must not be exercised unreasonably or vexatiously (cl 6.2.1). For example, for a breach of the CDM Regulations to merit termination, it should be sufficiently serious so as to cause a significant risk to health and safety and/or a risk of prosecution by the Health and Safety Executive. It should be noted that the defaults must occur 'without reasonable cause' on the contractor's part, and that, depending on the circumstances, the contractor might find 'reasonable cause' in any of the matters referred to in clause 6.8.1. An exercise of the right to suspend work under the HGCRA 1996 (as amended) would not be cause for termination, provided that it had been exercised in accordance with the terms of the contract.

9.11 One of the defaults listed in clause 6.4.1.2 is that the contractor 'fails to proceed regularly and diligently'. This is notoriously difficult to establish, and although meticulous records will help, contract administrators are often understandably reluctant to issue the first warning notice. Reported cases show how difficult this can be in practice. In the case of *London Borough of Hounslow* v *Twickenham Garden Developments*, for example, the architect's notice was strongly attacked by the defendant. In a more recent case, however, the architect was found negligent because he failed to issue a notice (*West Faulkner Associates* v *London Borough of Newham*). It should be remembered that if the contract administrator has not issued the first 'warning notice', the employer cannot issue the termination notice.

London Borough of Hounslow v *Twickenham Garden Developments* (1970) 7 BLR 81

The London Borough of Hounslow entered into a contract with Twickenham Garden Developments to carry out sub-structure works at Heston and Isleworth in Middlesex. The contract was on JCT63. Work on the contract stopped for approximately eight months due to a strike. After work resumed, the architects issued a notice of default stating that the contractor had failed to proceed regularly and diligently and that, unless there was an appreciable improvement, the contract would be determined. The employer then proceeded to determine the contractor's employment. The contractor disputed the validity of the notices and the determination, and refused to stop work and leave the site. The Borough applied to the court for an injunction to remove the contractor. The judge emphasised that an injunction was a serious remedy and that before he could grant one there had to be clear and indisputable evidence of the merits of the Borough's case. The evidence put before him, which showed a significant drop in the amounts of monthly certificates and numbers of workers on site, failed to provide this.

West Faulkner Associates v *London Borough of Newham* (1992) 61 BLR 81

West Faulkner Associates were architects engaged by the Borough for the refurbishment of a housing estate consisting of several blocks of flats. The residents of the estate were evacuated from their flats in stages to make way for the contractor, Moss, which, it had been agreed, would carry out the work according to a programme of phased possession and completion, with each block taking nine weeks. Moss fell behind the programme almost immediately. However, Moss had a large workforce on the site and continually promised to revise its programme and working methods to address the problems of lateness, poor quality work and unsafe working practices that were drawn to its attention on numerous occasions by the architect. In reality, Moss remained completely disorganised, and there was no apparent improvement. The architects took the advice of quantity surveyors that the grounds of failing to proceed regularly and diligently would be difficult to prove, and decided not to issue a notice. As a consequence, the Borough was unable to issue a notice of determination, had to negotiate a settlement with the contractor and dismissed the architect, which then brought a claim for its fees.

The judge decided that the architect was in breach of contract in failing to give proper consideration to the use of the determination provisions. In his judgment, he stated that 'regularly and diligently' should be construed together and in essence they mean simply that the contractors must go about their work in such a way as to achieve their contractual obligations. 'This requires them to plan their work, to lead and manage their workforce, to provide sufficient and proper materials and to employ competent tradesmen, so that the Works are carried out to an acceptable standard and that all time, sequence and other provisions are fulfilled' (Judge Newey at page 139).

Insolvency of the contractor

9.12 The term 'insolvent' is defined in the form at clause 6.1. In general terms, insolvency is the inability to pay debts as they become due for payment. Insolvent individuals may be declared bankrupt. Insolvent companies may be dealt with in a number of ways, depending upon the circumstances: for example, by voluntary liquidation (in which the company resolves to wind itself up), compulsory liquidation (under which the company is wound up by a court order), administrative receivership (a procedure to assist the rescue of a company under appointed receivers), an administration order (a court order given in response to a petition, again with the aim of rescue rather than liquidation, and managed by an appointed receiver), or voluntary arrangement (in which the company agrees terms with creditors over payment of debts). Procedures for dealing with insolvency are mainly subject to the Insolvency Act 1986 and the Insolvency Rules. Under these provisions, the person authorised to oversee statutory insolvency procedures is termed an 'insolvency practitioner'.

9.13 Under MW16 (cl 6.5.1), the employer has the right to terminate the contractor's employment in the event that the contractor is 'insolvent' as defined under clause 6.1. The contractual definition covers a wide range of situations, including voluntary arrangements and winding-up or bankruptcy orders. There is no requirement for the contractor to notify the employer in writing in the event of liquidation or insolvency, but as it is likely that such a requirement would be implied, the contractor might be expected to do this.

9.14 As termination is not automatic, the employer has, in effect, an option to consider a more constructive approach. If this is found not to be practicable then the only option will be for the employer to terminate the contractor's employment, and the only way to achieve completion will be by a new contractor of the employer's choice. In some circumstances, however, it may be sensible to allow an appointed insolvency practitioner time to come up with a rescue package. It is usually in the employer's interest to have the works completed with as little additional delay and cost as possible, and a breathing space might allow all possibilities to be explored. It may, for example, be possible for the contractor to continue and complete the works, provided funding can be arranged.

9.15 If the original contractor cannot continue, another contractor may be novated to complete the works. On a 'true novation', the substitute contractor takes over all the original obligations and benefits (including completion to time and within the contract sum). More likely is a 'conditional novation', whereby the contract completion date, etc. would be subject to renegotiation, and the substitute contractor would probably want to disclaim liability for that part of the work undertaken by the original contractor.

9.16 Deciding on which of the options would best serve the interests of all the parties is a matter for the employer, perhaps with the advice of the contract administrator and the insolvency practitioner. There might be a straightforward way out, or there might be advantages in taking a more pragmatic approach. For example, it may prove expeditious to continue initially with the original contractor under some interim arrangement until such time as novation can be arranged, or a completion contract negotiated.

9.17 Following termination, the employer may employ others to complete the work and may use any temporary buildings, equipment, etc. on the site for that purpose (cl 6.7.1). Clause 6.7.2 then states that 'no further sums shall become due to the Contractor under this Contract other than any amount that may become due to him under clause 6.7.4 and the Employer need not pay any sum that has already become due', which in effect means that other provisions of the contract requiring further payment will cease to operate. This relief from the obligation to make payments already due applies in only two circumstances: if the employer has already issued a pay less notice (cl 6.7.2.1) or if the contractor has become insolvent after the last date by which a pay less notice could have been issued (cl 6.7.2.2; this clause reflects section 110(10) of the HGCRA (as amended)). Therefore, if the contractor becomes insolvent after a certificate is issued, but before the final date for issuing a pay less notice, then to avoid any dispute the employer should issue a notice.

9.18 Once the works have been completed and any defects in them made good (within three months) an account is prepared of the cost of completing the work and the loss and/or expense and damages suffered as a result of the termination, offset against the amount that would have been paid under the contract (cl 6.7.3). The account is included in a certificate of the contract administrator, or statement of the employer, and the balance claimed from the contractor as a debt, or paid to the contractor as appropriate (cl 6.7.4).

Termination by the contractor

9.19 The contractor is also given the right to terminate its own employment if the employer defaults by failing to pay amounts due (cl 6.8.1.1), or attempts to interfere with or obstruct the issue of any certificate (cl 6.8.1.2) or fails to comply with the requirements of the CDM Regulations (cl 6.8.1.3). It should be noted that despite the wording of clause 6.8.1, which recognises 'the default or defaults', a single instance of late payment is unlikely to warrant termination since the contract contains a right of suspension. In addition, termination may be initiated if the works are suspended for a period of one month or more as a result of a contract administrator's instruction correcting an inconsistency or requiring a variation (cl 6.8.2.1), or because of a default or impediment by the employer, the contract administrator or any employer's person (cl 6.8.2.2), provided that the instruction or default was not necessitated by a default of the contractor. If the employer suspends the work, this default must affect the whole or 'substantially the whole' of the works for a continuous period of a month or more. In the event of the employer's insolvency, there is an option to terminate, but this is not automatic (cl 6.9). The contract expressly states that the clause 6.8 and 6.9 provisions for termination by the contractor do not prevent the exercise of other rights and remedies which it may possess (cl 6.3.1).

9.20 In the event of a clause 6.8 default, if the contractor wishes to terminate its employment it must first issue a notice, which must specify the default complained of and require it to be ended (cl 6.8.2). This is a warning notice to be sent to the employer, and the service of the notice must be in accordance with clause 6.2. If the default is not ended within seven days of the receipt of the notice, the contractor may then by further notice, or within ten days from the expiry of that seven-day period, terminate its employment. Termination then takes effect on the date of receipt of the notice. It appears that if termination did not result and the contractor was to repeat the default at some later time, then it might be necessary to go through the whole procedure of notices again.

9.21 Upon termination the contractor prepares an account setting out the total value of the work at the date of termination plus other costs relating to the termination, as set out in clause 6.11. These may include such items as the cost of removal and any direct loss and/or expense consequent upon termination. The contractor is in effect indemnified against any damages that may be caused as a result of the termination. This would not necessarily be the case if the contractor did not comply with the contractual provisions; in that case it might constitute repudiation. This account is then submitted to the employer, and the employer must pay the amount properly due within 28 days of its submission (cl 6.11.4).

Termination by either party

9.22 Either party is entitled to initiate termination if the works are suspended for a continuous period of one month or more due to various 'neutral' events as set out in clause 6.10.1. (This is in addition to the right under clause 5.7, which does not require a period of suspension; see paragraph 8.25). If a party wishes to implement this provision, it must issue a notice to that effect. If the suspension does not cease within a further seven days then that party may issue a second notice terminating the contractor's employment (cl 6.10.1). One of the listed 'neutral' events is loss or damage caused by any risk covered by the works insurance policy or by an excepted risk, and for this event the contract states that the contractor may not issue a notice where the damage has been caused by the negligence of the contractor or a contractor's person. Where the employer is a local or public authority, the employer may issue a notice if circumstances in section 73(1)(a) and (c) of the Public Contracts Regulations 2015 (relates to the awarding of contracts) apply (cl 6.10.3). Upon termination under clause 6.10, the provisions of clause 6.11 apply as described above, except that amounts of loss and/or damage caused to the contractor are not to be included in the statement of account. In effect the contractor bears the risk of such losses in the event of termination due to neutral events.

10 Dispute resolution

10.1 MW16 lists negotiation, mediation, adjudication, arbitration and legal proceedings as the means by which any disputes may be determined. The parties are required to decide in advance which of the processes will be used and make relevant deletions to the articles. It is important for the contract administrator to understand the options and to be able to give appropriate advice.

10.2 Article 6 states the right of each party to refer 'any dispute or difference' to adjudication. This right comes, of course, from the Housing Grants, Construction and Regeneration Act (HGCRA) 1996 (as amended), which applies to all 'construction contracts'. MW16, however, may well be used in situations which come within the exception set out in section 106 of the HGCRA 1996, which states that the Act does not apply to a 'construction contract with a residential occupier'. In such cases there will be no statutory right to adjudication, but, unless the parties indicate otherwise, adjudication will still be a contractual right. The wording of the contract would have to be amended if this is not to apply. The parties should give consideration to the relative merits of adjudication, arbitration and litigation, in particular whether they wish an immediate albeit 'temporary' solution, or whether they would prefer a final decision, such as that which would be provided by short-form arbitration. In any event, the form should never be amended unless the parties are quite certain that the Act does not apply to their contract.

10.3 Where Article 7 applies, final determination of any dispute or difference will be by means of arbitration, and this will be in accordance with Schedule 1 and the JCT 2016 edition of the Construction Industry Model Arbitration Rules (CIMAR). If the parties prefer to use arbitration, this must be indicated in the contract particulars, otherwise the dispute will be resolved by litigation, then Article 7 will be deleted and Article 8 becomes operative. It is important that the contract administrator explains the implications of arbitration and, if the employer is a consumer, the possible effects of Part 2 (unfair terms) of the Consumer Rights Act 2015. It should be noted that the arbitration agreement may be considered 'unfair' with respect to that Part in so far as it relates to disputes involving small sums, as provided by section 91 of the Arbitration Act 1996. Before advising on this decision, the contract administrator should also establish whether the employer is likely to be eligible for Legal Aid. If Legal Aid is a possibility, then it would undoubtedly be better to select litigation.

Alternative dispute resolution

10.4 A better initial approach might be to embark on negotiations or to adopt some voluntary method of agreement before formal procedures are invoked. Initially the parties may try negotiation. If Supplemental Provision 6 is included, this sets out a procedure to be followed, whereby parties must notify each other as soon as a potential dispute emerges and then meet to try to resolve it. The contract administrator should tread carefully before becoming involved. Advising the client and providing information might be of great help, but the contract administrator will usually have no authority to negotiate terms or agree to contract amendments. Even if authority is given to negotiate a settlement, the architect should take care not to be drawn into complex areas of law which might be better left to a lawyer to handle.

10.5 MW16 includes mediation under clause 7.1, whereby the parties are required to give serious consideration to any request by either party to use mediation. This is not an absolute requirement, but a party should not dismiss a request out of hand, and if it would prefer not to mediate, should be clear as to its reasons for refusal.

10.6 Nevertheless, there can be many advantages to mediation. Unlike adjudication, arbitration or litigation, it is a non-adversarial process which tends to forge good relationships between the parties. Imposed solutions may leave at least one of the parties dissatisfied and may make it very difficult to work together in the future. If the parties are keen to promote a long-term business relationship they should give mediation serious consideration. Even if mediation does not result in a complete solution, it has been found in practice that it can help to clear the air on some of the issues involved and to establish common ground. This, in turn, might then pave the way for shorter and possibly less acrimonious arbitration or litigation.

Adjudication

10.7 The HGCRA 1996 Part II (as amended) requires that parties to the construction contracts falling within the definition set out in the Act have the right to refer any dispute to a process of adjudication which complies with requirements stipulated in the Act. Article 6 of MW16 restates this right, and refers to clause 7.2, which states that where a party decides to exercise this right 'the Scheme shall apply'. This refers to the Scheme for Construction Contracts (the Scheme), a piece of secondary legislation which sets out a procedure for the appointment of the adjudicator and the conduct of the adjudication. The Scheme takes effect as implied terms in a contract, if and to the extent that the parties have failed to agree on a procedure that complies with the Act.

10.8 By stating 'the Scheme shall apply', MW16 is effectively annexing the provisions of the Scheme to the form, which therefore become a binding part of the agreement between the parties. Clause 7.2 allows for the parties to determine how the adjudicator will be appointed, by means of appropriate insertions in the contract particulars. Under MW16 the adjudicator may either be named in the contract particulars, or nominated by the nominating body identified in the contract particulars. It should be noted that the list includes, in addition to the RIBA, the RICS and the CIArb, the 'constructionadjudicators.com' and the Association of Independent Construction Adjudicators (AICA).

10.9 The party wishing to refer a dispute to adjudication must first give notice under paragraph 1(1) of the Scheme (see Figure 10.1). The notice may be issued at any time and should identify briefly the dispute or difference, give details of where and when it has arisen, set out the nature of the redress sought, and include the names and addresses of the parties, including any specified for the giving of notices (Scheme: paragraph 1(3)). If the adjudicator is named in the contract, he or she will normally have already signed terms of appointment on the JCT Adjudication Agreement for a Named Adjudicator (Adj/N). If no adjudicator is named, the parties may either agree an adjudicator or either party may apply to the 'nominator' identified in the contract particulars (paragraph 2(1)). If no nominator has been

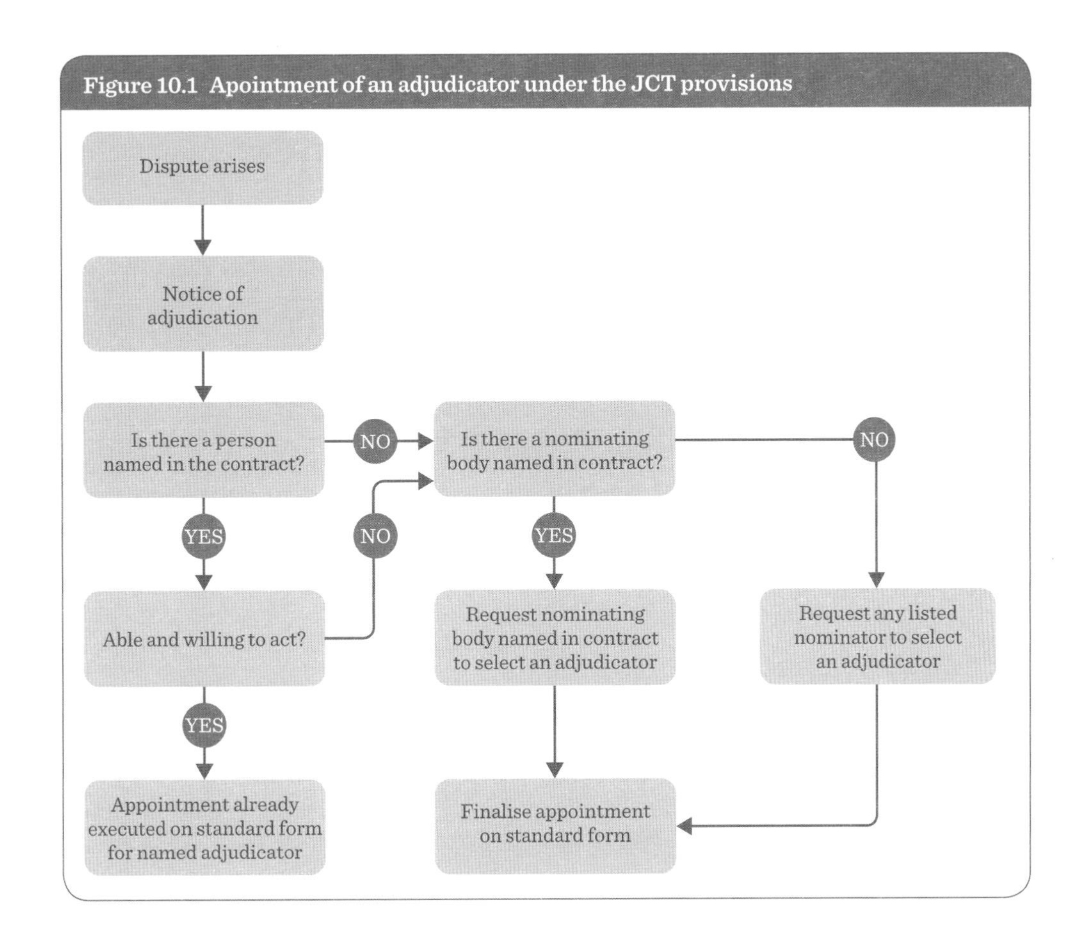

Figure 10.1 Apointment of an adjudicator under the JCT provisions

selected, then the contract states that the referring party may apply to any of the nominators listed in the contract particulars. The adjudicator will then send terms of appointment to the parties. In addition to the form for a named adjudicator, the JCT also publishes an Adjudication Agreement (Adj) for use in this situation.

10.10 The Scheme does not stipulate any qualifications in order to be an adjudicator, but does state that the adjudicator 'should be a natural person acting in his personal capacity' and should not be an employee of either of the parties (paragraph 4). If the adjudicator is to be named, the subject matter of the dispute will not be known at that time, so it would be sensible to name an adjudicator with a broad range of experience. If the nominating body route is used, that body will normally select an adjudicator with suitable experience from its panel. Where the person appointed does not have the appropriate expertise, that person must appoint an independent expert to advise and report. The adjudicator is required to act impartially, and to avoid unnecessary expense (paragraph 12).

10.11 The referring party must refer the dispute to the selected adjudicator within seven days of the date of the notice (paragraph 7(1)). The referral will normally include particulars of the dispute, a summary of the contentions on which the referring party relies, a statement of relief or remedy sought and any material the referring party wishes the adjudicator to consider, and must include a copy of, or relevant extracts from, the contract (paragraph 7(2)).

A copy of the referral must be sent to the other party, and the adjudicator must inform all parties of the date it was received (paragraph 7(3)).

10.12 The adjudicator will then set out the procedure to be followed. A preliminary meeting may be held to discuss this, otherwise the adjudicator may send the procedure and timetable to both parties. The party which did not initiate the adjudication (the responding party) will be required to respond, normally within seven days of the date of referral. The adjudicator may hold a short hearing at which the parties can put forward further arguments and evidence; however, it is common for a dispute to be dealt with by 'documents only', i.e. the adjudicator does not meet with the parties at any point.

10.13 The adjudicator is given considerable powers under the Scheme (e.g. paragraphs 13 and 20), including the right to take the initiative in ascertaining the facts and the law, the right to issue directions, the right to revise decisions and certificates of the contract administrator, the right carry out tests (subject to obtaining necessary consents) and the right to obtain from others necessary information and advice. The adjudicator must give advance notice if intending to take legal or technical advice (paragraph 13(vi)).

10.14 The HGCRA 1996 (as amended) requires that the decision is reached within 28 days of referral, but it does not state how this date is to be established (section 108(2)(c)). Under the Scheme, the 28 days start to run from the date of receipt of the referral notice (paragraph 19(1)). The period can be extended by up to 14 days by the referring party, and further by agreement between the parties. The decision must be delivered forthwith to the parties, and the adjudicator may not retain it pending payment of the fee. The provisions state that the adjudicator must give reasons for the decision if requested to do so by the parties (paragraph 22).

10.15 The parties must meet their own costs of the adjudication, unless they have agreed that the adjudicator shall have the power to award costs. Under the Act, any agreement is ineffective unless it complies with section 108A, including that it is made in writing after a notice of adjudication is issued (MW16 therefore does not contain such an agreement). The adjudicator, however, is entitled to charge fees and expenses (subject to any agreement to the contrary), although expenses are limited to those 'reasonably incurred' (paragraph 25). The adjudicator may apportion those fees between the parties, and the parties are jointly and severally liable to the adjudicator for any sum which remains outstanding following the adjudicator's determination. This means that, in the event of default by one party, the other party becomes liable to the adjudicator for the outstanding amount.

10.16 The adjudicator's decision will be final and binding on the parties 'until the dispute is finally determined by legal proceedings, by arbitration, or by agreement between the parties'. The effect of this is that if either party is dissatisfied with the decision, it may raise the dispute again in arbitration or litigation, or may negotiate a fresh agreement with the other party. If the dispute is raised again in a further tribunal, the dispute would be considered again from scratch, with new evidence if necessary, and would not be in the form of an appeal against the adjudicator's decision. In all cases, however, the parties remain bound by the decision and must comply with it until the final outcome is determined.

10.17 If either party refuses to comply with the decision, the other may seek to enforce it through the courts. Generally, actions regarding adjudicators' decisions have been dealt with promptly and firmly by the courts and the recalcitrant party has been required to comply.

Paragraph 22A of the Scheme allows the adjudicator to correct clerical or typographical errors in the decision, within five days of it being issued, either on the adjudicator's own initiative or because the parties have requested it, but this would not extend to reconsidering the substance of the dispute.

Arbitration

10.18 Arbitration refers to proceedings in which the arbitrator has power derived from a written agreement between the parties to a contract, and which is subject to the provisions of the Arbitration Act 1996. Arbitration awards are enforceable at law. An arbitrator's award can be subject to appeal on limited grounds.

10.19 If arbitration is selected as the method for final determination of disputes, then this is confirmed by making the appropriate deletion in the contract particulars. The arbitration provisions are referred to in clause 7.3 and set out in Schedule 1: Arbitration, which refers to the Construction Industry Model Arbitration Rules ('the Rules'). The Arbitration Act 1996 confers wide powers on the arbitrator unless the parties have agreed otherwise, but leaves detailed procedural matters to be agreed between the parties or, if not so agreed, to be decided by the arbitrator. To avoid problems arising, it is advisable to agree as much as possible of the procedural matters in advance, and MW16 does this by incorporating the Rules, which are very clearly written and self-explanatory. The specific edition referred to is the 2016 edition of CIMAR published by the JCT (Article 7), which incorporates supplementary and advisory procedures, some of which are mandatory (Part A) and some of which only apply if agreed after the arbitration is begun (Part B). The paragraphs below refer to the JCT edition of the Rules.

10.20 The party wishing to refer the dispute to arbitration must give notice as required by MW16 Schedule 1 paragraph 2.1 and Rule 2.1, briefly identifying the dispute and requiring the other party to agree to the appointment of an arbitrator. If the parties fail to agree within 14 days, either party may apply to the 'appointor', selected in advance from a list of organisations set out in the contract particulars. If no appointor is selected, then the contract states that the appointer shall be the president or a vice-president of the RIBA.

10.21 The arbitrator has the right and the duty to decide all procedural matters, subject to the parties' right to agree any matter (Rule 5.1). Within 14 days of appointment the parties must each send the arbitrator and each other a note indicating the nature of the dispute and amounts in issue, the estimated length for the hearing and the procedures to be followed (Rule 6.2). The arbitrator must hold a preliminary meeting within 21 days of appointment to discuss these matters (Rule 6.3). The first decision to make is whether Rule 7 (short hearing), Rule 8 (documents only) or Rule 9 (full procedure) is to apply. The decision will depend on the scale and type of dispute.

10.22 Under all three Rules referred to above, the parties exchange statements of claim and of defence, together with copies of documents and witness statements on which they intend to rely. Under Rule 8, the arbitrator makes an award based on documentary evidence only. Under Rule 9, the arbitrator will hold a hearing at which the parties or their representatives can put forward further arguments and evidence. There may also be a site visit. The JCT amendments set out time limits for these procedures.

10.23 Under Rule 7 a hearing is to be held within 21 days of the date when Rule 7 becomes applicable, and the parties must exchange documents not later than seven days prior to the hearing. The hearing should last no longer than one day. The arbitrator publishes the award within one month of the hearing. The parties bear their own costs.

10.24 The arbitrator is given a wide range of powers under Rule 4, including the power to obtain advice (Rule 4.2), the powers set out in section 38 of the Arbitration Act 1996 (Rule 4.3), the power to order the preservation of work, goods and materials even though they are a part of work that is continuing (Rule 4.4), the power to request the parties to carry out tests (Rule 4.5) and the power to award costs. Under Schedule 1 paragraph 3 of MW16 the arbitrator is also given wide powers to review and revise any certificate, opinion, decision, requirement or notice, and to disregard them if need be, where seeking to determine all matters in dispute.

10.25 Where the arbitrator has the power to award costs, this will normally be done on a judicial basis, i.e. the loser will pay the winner's costs (Rule 13.1). The arbitrator will be entitled to charge fees and expenses and will apportion those fees between the parties on the same basis. The arbitrator's terms of appointment will normally state that the parties are jointly and severally liable to the arbitrator for fees and expenses incurred.

Litigation

10.26 Where the parties have rejected arbitration and deleted Article 7, then Article 8 becomes operative and disputes will be determined by legal proceedings.

10.27 Proceedings are usually initiated by the claimant filing a claim at the appropriate county court (this can be done online if the amount of the claim is known). The court will then allocate the case to a 'track' (small claims, fast track or multi track) depending on its size and complexity. Larger and more complex construction cases are usually heard in the Technology and Construction Court, a specialist department of the High Court which deals with technical or scientific cases. Procedures in court follow the Civil Procedure Rules, with the timetable and other detailed arrangements being determined by the court. A judge will hear the case, and although, in the past, parties were required to be represented by barristers, now they may represent themselves, or elect to be represented by an 'advisor'.

10.28 Disputes in building contracts have traditionally been settled by arbitration. Arbitrators are usually senior and experienced members of one of the construction professions, and for many years it was felt that they had a greater understanding of construction projects and the disputes that arise than might be found in the courts. These days, however, the judges of the Technology and Construction Court have extensive experience of technical construction disputes. The high standards now evident in these courts are likely to be matched in practice by only a few arbitrators.

10.29 The court has powers to order that actions regarding related matters are joined (for example, where disputes between an employer and contractor, and contractor and sub-contractor, concern the same issues). This is much more difficult to achieve in arbitration. Even if all parties have agreed to the use of CIMAR, the appointing bodies must have been alerted and have agreed to appoint the same arbitrator (Rules 2.6 and 2.7). If the same arbitrator is appointed, he or she may order concurrent hearings (Rule 3.73), but may only order consolidated proceedings with the consent of all the parties (Rule 3.9), which is often difficult to obtain. The court's powers may therefore offer an advantage in multi-party disputes, by avoiding duplication of hearings and the possibility of conflicting outcomes.

10.30 There remain, however, two key advantages to using arbitration. The first is that in arbitration the proceedings can be kept private, which is usually of paramount importance to construction professionals and firms, and is often a deciding factor in selecting arbitration. In court, the proceedings are open to the public and the press, and the judgment is published and widely available.

10.31 The second advantage to the parties is that the arbitration process is consensual. The parties are free to agree on timing, place, representation and the individual arbitrator. This autonomy carries with it the benefits of increased convenience, and possible savings in time and expense. The parties avoid having to wait their turn at the High Court, and may choose a time and place for the hearing which is convenient to all. In arbitration, however, the parties have to pay the arbitrator and meet the cost of renting the premises in which the hearing is held.

10.32 Where parties have selected arbitration under Article 7, it is still open for them to select litigation once a dispute develops. If, however, one party commences court proceedings, the other may ask the court to stay the proceedings on the grounds that an arbitration agreement already exists. This would not apply to litigation to enforce an adjudicator's decision, as Article 7 excludes all disputes regarding the enforcement of a decision of an adjudicator from the jurisdiction of the arbitrator.

References

Publications

Chappell, D. *The JCT Minor Works Building Contracts 2005*, Blackwell Publishing, Oxford (2006).
Finch, R. *NBS Guide to Tendering: For Construction Projects.* RIBA Publishing, London (2011).
Furst, S. and Ramsey, V. (eds) *Keating on Construction Contracts*, Sweet & Maxwell, London (2006).
JCT. *Tendering Practice Note 2012.* Sweet & Maxwell, London (2012).
Ostime, N. *RIBA Job Book*, 9th edition, RIBA Publishing, London (2013).
RICS. *New Rules of Measurement: Detailed Measurement for Building Works*, RICS, London (2012).
RICS and Davis Langdon. *Contracts in Use: A Survey of Building Contracts in Use During 2010*, RICS, London (2012).

Cases

Legislation

Statutes

Statutory instruments

Clause Index *for MW16 by paragraph*

Clause Index *for MWD16 by paragraph*

Subject Index *by paragraph*